# Principles of Plant Breeding

# Principles of Plant Breeding
## Second Edition

ROBERT W. ALLARD

JOHN WILEY & SONS, INC.

New York / Chichester / Weinheim / Brisbane / Singapore / Toronto

This book is printed on acid-free paper. ∞

*Library of Congress Cataloging-in-Publication Data:*

Allard, R. W. (Robert Wayne), 1919–
    Principles of plant breeding / Robert W. Allard. — 2nd ed.
      p.  cm.
    ISBN 0-471-02309-4 (cloth : alk. paper)
    1. Plant breeding. I. Title.
  SB123.A5  1999
    631.5'2—dc21                                             98-47144

10 9 8 7 6 5 4 3 2 1

# *Contents*

# *Preface*

Humans are dependent on plants for survival. Photosynthesizing plants are the initial source of human sustenance and the original source, directly or indirectly, of most clothing, fuel, construction materials, and medicinals. Moreover, as ornamentals, they are both useful and esthetically pleasing. Considering the major importance of plants, it is not surprising that humans have, since plants were first domesticated about 13,000 to 11,000 years ago, been energetically involved in developing plants, especially food plants, that better serve their needs. Only recently have these attempts been systematized to the point that they can be called a science. This quite new science, the science of plant breeding, is the subject of this book.

The approach to plant breeding that was ultimately adopted in this, the second edition of *Principles of Plant Breeding*, evolved gradually in response to queries put to many colleagues in various disciplines, especially biologists, regarding those aspects of science and technology that have come to be regarded as most important in developing potentially more useful plants for the present world. In addressing the many suggestions received, this edition of *Principles of Plant Breeding* starts with three chapters based largely on the central ideas of Charles Darwin's *Origin of Species* (1859). Devoting three early chapters largely to Darwin, however, in no way belittles the major contributions made by domesticators who applied important selection pressures on the wild plants they chose to bring into cultivation. Nor does it deprecate the contributions of early agriculturalists whose selective preferences greatly modified early domesticates as they spread locally and later were dispersed, some throughout much of the world, often into environmentally quite different ecological situations. Its purpose is to recognize (1) the major roles subsequently played by Darwin's ideas and (2) the major support that the carefully documented changes that have occurred in cultivation described by de Candolle (1866) and many others have provided for Darwin's ideas. Part I culminates in two chapters that address ideas of central importance in plant breeding: Chapter 4, which focuses on the mating and reproductive systems that determine the methods of breeding appropriate to three main groups of plants (self-pollinating plants, cross-pollinating plants, plants that can be propagated vegetatively), and Chapter 5, which surveys the major effects of mating systems, genetic linkage, and epistasis have on the improvement of cultivated plants.

Part II (6 chapters) focuses on the genetic principles on which all breeding plans are based. First, the core ideas of Mendelian inheritance are presented in Chapter 6, followed by descriptions in Chapter 7 of the impressive powers of Mendelian segregation and recombination in generating enormous amounts of heritable variability. The extraordinary complexities of continuously varying characters and their analysis with the use of biometrical methods are then considered (Chapter 8), followed by an example (Chapter 9) of the intricacies of long-term selection (hundreds of generations) for improved adaptedness during the evolution of cultivated corn. The two final chapters of Part II (Chapters 10 and 11) illustrate a very recent development, namely, the use of discretely inherited marker alleles to determine more precisely what happened genetically during the evolution of adaptedness in nature as well as in cultivation.

Part III (Chapters 12 through 17) is concerned with the development and operational features of modern breeding plans. Chapter 12 focuses on the overwhelming influence of mating and/or reproductive systems (selfing, outcrossing, clonal reproduction) in determining the breeding plans appropriate to particular species. Details of the development and operational features of breeding plans that are presently applied to self-pollinated, outcrossing, and clonally reproducible species, and the breeding of hybrid varieties of selfing and clonally propagated species, are considered in Chapters 13, 14, 15, and 16 respectively.

Chapter 17 addresses problems that have attracted much attention internationally during recent years. Particularly vexatious are certain features of genotype-environment interactions that have proved troublesome in the many areas of the world in which agriculture is practiced by resource-poor human populations. It has been estimated that about 1.4 billion people worldwide depend on subsistence agricultures and that, even though resource-poor farmers conduct about 60% of the total global agriculture, they produce less than 20% of food globally. Unfortunately, breeding programs have typically been less efficient in resource-poor than in resource-rich areas, often because genotype $\times$ environment interactions frequently lead to serious problems, particularly in subsistence agricultures, in which agricultural conditions are often harsh and erratically unpredictable. Recently, methods have been developed that more accurately identify appropriate matches between genotypes and environments. This joint genotype-environment concept is proving useful, because planting the winning, genotypes in each environment maximizes total yield. This concept is the main focus of Chapter 17.

It seems appropriate to end this preface by expressing my thanks to the many people—farmers, seedsmen, plant and animal breeders, geneticists, ecologists, and other professionals from many nations—who, in discussions over the years, helped immensely in clarifying my ideas as to what is important in plant breeding. These contributors are too numerous to mention individually. However, I would indeed be remiss if I did not acknowledge the guidance of my father (a skilled farmer-seedsman) and W. W. Mackie, University of California, Berkeley, who conducted experiments on my parents' farm during my primary and secondary school years. During this period Professor Mackie played a major role in starting my scientific career by introducing me to Darwin's *Origin of Species*, to Gregor Mendel's *Versuche über Pflanzenhybriden*, and to the English philosophers, ranging from Bacon to Mill. It is also appropriate to acknowledge my gratitude to numerous colleagues, technicians, support staff, undergraduate and graduate students, and faculty of various departments of the University of California, both Davis and Berkeley, during four undergraduate years and the more than forty years of my career on the faculty at the University of California. Graduate studies at the University of Wisconsin and sabbatical leaves at various universities in South America, Europe, the Middle East, and Asia also broadened my perspectives. Finally, I thank Ann Allard for word processing and editing this second edition of *Principles of Plant Breeding*, as well as her much appreciated support and understanding during my entire scientific career.

R. W. Allard

Professor Emeritus of Genetics and of Agronomy and Range Science

University of California, Davis

Bodega Bay, California

August 1998

# PART I

*Introductory Topics*

# ONE

# Darwinian Evolution

In broadest terms, this second edition of *Principles of Plant Breeding* is concerned with the two distinct components of Darwinian evolution: (1) the facts of evolution ("descent with modification") and (2) selection as the chief agent of evolutionary change. However, the narrower focus of the book is on two quite recent and specialized evolutionary events: first, the evolution of crop plants under cultivation, a process that started about 13,000 to 11,000 years ago with the cultivation and ultimate domestication of many wild plant species in various parts of the world; and second, the current scientific phase of plant breeding, a process that started early in the twentieth century as Darwinian and Mendelian principles became firmly established as the framework within which evolutionary changes in living organisms occur. Because life as we know it has been conditioned by planet Earth, it seems appropriate to introduce *Principles of Plant Breeding* with a brief description of the grand scale of the evolution of Earth, from the formation of the planets of the solar system, to the origin of life on Earth, through biological evolution from the earliest primal living systems, to the evolution of modern plants during domestication and cultivation.

Current cosmological opinion holds that the universe started perhaps 10 billion ($10 \times 10^9$) years ago in a violent explosion, the "big bang." One plausible hypothesis is that cosmic debris, from the big bang from which the solar system was formed, was originally distributed as a cloud of gases, dust, and many larger particles, scattered thinly throughout a volume probably somewhat larger than that of the present solar system. As this cloud collapsed under gravity, it spun faster and faster until the rotation caused the cloud to flatten into a disk, with the greatest density at the center. About the same time that the center became sufficiently massive to support a nuclear fusion reaction, the remains of the rotating disk presumably started clumping together to form the nine planets of the solar system and their satellites. Some celestial bodies, with sizes varying from specks of dust to asteroids weighing many tons, still orbit the sun. The ~ $4\frac{1}{2}$ billion-year age assigned the Earth presumably reflects the time when the aggregation of various sized particles into a planet was nearly complete. Evidently, the Earth and meteors formed at about the same time; if this was indeed the case, the age of both can be dated from the radiometric ages of the most ancient materials of the Earth and meteorites.

During the original aggregation, planet Earth presumably lost most of the gases not contained within solid materials because gravitational attraction was inadequate to prevent lightweight gas molecules from escaping into space. But when the planet approached its present size, heating of its interior caused water and gases frozen on the solid chunks of matter to evaporate. These lighter materials then apparently rose to the surface, but gravity would at some point have become adequate to prevent their escape. The early atmosphere of Earth thus gradually accumulated, but mostly as secondary products derived from the Earth's interior. Probably a major condensate during this process was water, which rose to the surface to form primordial shallow seas. Then, as now, oceans covered most of the globe, a grayish expanse broken by small continents and volcanic platforms that rose out of the then shallow oceans. Sediments show that the chemical content of seawater at that time was more influenced by circulation of water through vents in the ocean floor than by erosion from the land, as is presently the case. Many geologists accept the notion that the major flow of lighter materials into the atmosphere and hydrosphere, and heavier materials into the Earth's crust, mantle, and core, occurred within a few million years after the Earth had assumed much of its present form. The atmosphere at that time is believed to have contained abundant carbon dioxide, nitrogen, methane, and compounds containing sulfur and chlorine (gases that still escape from volcanic vents), but very little oxygen. It is almost certain that no modern living form could have survived more than briefly in such an environment.

There is no direct evidence concerning the origin of life. Only two kinds of direct evidence seem possible: (1) finding fossils of the first organisms or (2) producing living organisms from nonliving matter in the laboratory. Neither possibility seems likely, especially that of finding fossilized traces of the aggregations of molecules that preceded the first truly living organisms, probably a primitive order of bacterialike organisms (*Archaea*) with descendants that have survived to the present. The first living things were almost certainly very small, and their composition was such that they were most unlikely to have become fossilized, or if they had become fossilized, to have survived as recognizable fossils to the present. The origin of life thus remains a mystery, but a possible hypothesis is that life could and perhaps did arise from nonliving material by natural processes along lines proposed by Stuart Kauffman in his 1995 book *At Home in the Universe*. Underlying all present living systems is an intricate network of chemical reactions guided by enzymes: hence, it is most unlikely that the chemical reactions essential to all life could have arisen in the absence of enzymes. Enzymes are long, unbranched molecules made up of amino acid subunits. Amino acids can be formed by passing electric sparks through gases (e.g., water vapor, methane, ammonia) known to be present in the early atmosphere of the Earth; moreover, the conditions found in present-day high-temperature thermal vents in the ocean floor are known to produce complex organic compounds, including amino acids. At present there is only one group of early amino acids known that can copy themselves. These are two nucleic acids, ribonucleic acid (RNA) and deoxyribonucleic acid (DNA), and it is now considered likely these two nucleic acids somehow became jointly capable of self-reproduction in the sense that, given a suitable source of energy, they served as patterns (templates) by which they were able to replicate themselves. If this hypothesis is correct, the basic mechanism of heredity on Earth would have been in place soon after the planet had cooled sufficiently for life to exist and the planet had evolved to reach approximately its present form. Given the existence of substantial variations among the many kinds of molecules that were likely to have formed, it also seems likely that natural selection would have occurred and would have favored the more effective aggregates of molecules. The result might eventually have been multimolecular aggregates endowed with the basic attribute of life, namely, the ability to reproduce their own kind. According to this hypothesis, the raw materials from which the earliest self-reproducing multimolecular aggregates formed were very large numbers of organic compounds that arose from comparatively simple nonliving materials. Thus, there may be no precise point at which a dividing line can be drawn between the living and nonliving.

Understanding the history of life becomes less hypothetical when fossils are available. Direct evidence of early organisms is provided by fossils of true bacteria (*Eubacteria*) preserved in rocks about 3.8 billion years old. As noted previously, the earliest biota of the Earth probably arose in hot, oxygen-poor environments, utilizing energy from chemical reactions. Early on, one lineage of *Eubacteria*, the *Cyanobacteria*, developed the ability to use the abundant energy from sunlight to form organic matter from carbon dioxide dissolved in seawater. And because oxygen is a by-product of photosynthesis, an apparently entirely new course was set in the environmental history of Earth, opening the way for the evolution of organisms with aerobic oxygen-driven metabolisms. However, the oxygen-driven revolution was slow. *Cyanobacteria* may have begun releasing oxygen into the atmosphere more than 3.5 billion years ago, but perhaps more than an additional billion years passed before the atmosphere contained sufficient oxygen to allow the development of oxygen-dependent organisms. Still higher oxygen levels eventually developed (the present level is ~ 21%). Another biologically significant threshold was reached when the stratospheric ozone ($O_3$) layer had developed to the point where it provided an effective shield from damaging ultraviolet radiation from the Sun. The earliest photosynthetic organisms were almost certainly prokaryotes. Carbon atoms occur in two distinctive forms that differ by a single neutron. Photosynthetic organisms incorporate the lighter form; consequently, they can be identified by the type of carbon preserved in early rocks, giving direct evidence that complex aerobic microbial communities existed more than 3 billion years ago (see Table 1–1).

The Earth, then as now, was ever changing, and with every change disruptions, and often extinction, occurred for some forms of life. Yet change offered new opportunities, with the result that the numbers and kinds of environments available for colonization were greatly increased over time. The biological impact of the oxygen-enriched (oxygen-polluted) environment is evident in the evolutionary pathways that reveal the genealogical relationships between organisms. All of the earlier branches of the evolutionary tree are occupied by organisms that were unable to use oxygen in their metabolism; indeed, oxygen was toxic for many of these organisms. Species on all later branches of the evolutionary tree are able to use oxygen in respiration, the energy-yielding process by which organic molecules

TABLE 1-1   Geological time scale and history of life on Earth

| | |
|---|---|
| *Precambrian Era* | 4,600 to 600 million years ago. Extent, 4,000 million years. Fossils of earliest simple marine forms (bacteria, algae, invertebrates) appear. |
| *Paleozoic Era* | 600 to 225 million years ago. Extent, 375 million years. |
| | *Cambrian* Period. Extent, 120 million years. Abundant fossils, all marine. |
| | *Ordovician* Period. Extent, 45 million years. Early amphibians. |
| | *Silurian* Period. Extent, 35 million years. First vascular plants. |
| | *Devonian* Period. Extent, 50 million years. Diversification of vascular plants. |
| | *Carboniferous* Period. Extent, 90 million years. Rapid diversification of horsetails and club mosses. |
| | *Permian* Period. Extent, 35 million years. Rapid diversification of fernlike plants. |
| *Mesozoic Era* | 225 to 70 million years ago. Extent, 155 million years. |
| | *Triassic* Period. Extent, 45 million years. Rapid diversification of gymnosperms. First dinosaurs and mammals. |
| | *Jurassic* Period. Extent, 50 million years. Angiosperms and first birds appear. |
| | *Cretaceous* Period. Extent, 60 million years. Rapid diversification of monocots and dicots. Dinosaurs in ascendance. |
| *Cenozoic Era* | 70 million years ago to present. Extent, 70 million years. |
| | *Tertiary* Period. Extent, 67 million years. |
| |     *Paleocene* Epoch. Earliest large mammals. |
| |     *Eocene* Epoch. Earliest grasses. |
| |     *Oligocene* Epoch. |
| |     *Miocene* Epoch. Earliest hominids. |
| |     *Pliocene* Epoch. |
| | *Quartinary* Period |
| |     Pleistocene 1.6 million years. Earliest humans (*Homo*). |
| |     Holocene 15,000–10,000 years. Agriculture arises. |

Adapted from several sources, including Merriam Webster's *Collegiate Dictionary*, 10th edition, 1993, and *American Heritage Dictionary*, 1978.

are broken down into $CO_2$ and $H_2O$. The fossil record suggests that in the novel experiments created by the oxygen revolution, bacteria may have diversified into many aerobic lineages (the Cambrian explosion) during late Precambrian or early Cambrian times, whereas ancient anaerobic lineages became less abundant as their environments became more restricted. Nevertheless, anaerobic lineages have continued their role in cycling carbon and other elements up to the present.

The next huge step for life started about 440 million years ago with the colonization of the land in the Silurian Period of the Paleozoic Era. Perhaps the single most important adaptation for plants was the ability to produce a cuticle, a waxlike covering of the exterior surface. Cuticles are not required for life in water but are essential for protection from desiccation in the air. Development of stomata was another essential step, because aerobic plants require oxygen and $CO_2$ for respiration and photosynthesis. Guard cells that regulate the size of the openings in the stomata are also required to control the rate of loss of water vapor into the air. The oldest fossil plants with cuticles, stomata, and guard cells date back about 410 million years. Another important adaptation was the evolution of asexual reproductive cells (spores) with cutinized walls that reduce water loss. The earliest plants with cuticles, stomata, and cutinized spores date back about 400 million years.

This initial stage in the colonization of the land by plants was one of the most significant adaptive events in the history of life on Earth. Life on land must have posed great difficulties for aquatic organisms. It required giving up immersion in water, essential for growth and reproduction, as well as for coping with the effects of gravity on body mass (air is much less dense than water and offers little support for soft parts against the force of gravity). On the other hand, plants had much to gain by leaving the water. For example, because water absorbs light much more quickly than air—about half of the available energy in light is absorbed in the top centimeter of a body of water—photosynthesis is hence efficient in water only at very shallow depths.

The second major phase of land-plant evolution began with the appearance of cells that were capable of conducting sap throughout the entire plant body. Vascular tissues allowed large increases in plant size and efficiency. In addition, extensive root systems made possible tall, sturdy stems, capable of elevating leaves above the interference of smaller competitors. Starting about 400 million years ago, within the Silurian and Devonian Periods, four major plant groups (horsetails, club mosses, ferns, gymnosperms) appeared, all of which have survivors into the present. Two of these groups (horsetails and club mosses) underwent great taxonomic explosions during the next 220 million years (Carboniferous through Jurassic) but decreased in diversity thereafter, whereas ferns and, particularly, gymnosperms have increased in diversification up to the present. Flowering plants (monocots and dicots) made their first appearance in the fossil record about 125 million years ago (late in the Mesozoic Era); they diversified rapidly during the Cretaceous and have maintained dominance up to the present.

The preceding narrative indicates that the diversification of plant life was far from haphazard. Although documentation of the origin of the first life on Earth and the early history of the most simple organisms is sketchy, educated conjectures can be made about the progressions by which life evolved from nonliving materials. The fossil record provides more secure documentation of the evolutionary history of single-celled organisms and, still further, provides secure information concerning the relatively slow diversification of early multicellular organisms during the ~ 4,600 million years of the Precambrian. Rates of diversification were also modest during much of the Paleozoic (600 to 225 million years ago) but were much more rapid through most of the Mesozoic (225 to 70 million years ago) and Cenozoic (70 million years ago to present). Bursts of diversification leading to major new families usually coincided with the occupation of new habitats and the opportunities thus provided for development of novel adaptations. The new taxonomic groups were usually new clades (similar groups descended originally from a single ancestor). New clades often do not replace preexisting clades completely, which, thus, usually leads to increases in the overall diversity of life. However, extinctions also occurred as environmental conditions changed, resulting in periods of loss of diversity. Recent quantitative estimates of the diversity of all organisms indicate that diversity increased slowly in the very long periods of the Precambrian and Early Cambrian times, increased rapidly during the Cambrian explosion to a global value of about 280 families, fell to about 120 families in the late Cambrian, and then again increased during the Ordovician to about 450 families. Thereafter, diversity rose gradually to about 600 families by the end of the Paleozoic Era, fell to about 450 families during the early Mesozoic, again rose rapidly to about 1,260 families by the end of the Cretaceous, and then rose very rapidly to about 2,150 families during the Cenozoic Era.

Nearly all human (*Homo sapiens*) societies have speculated and created stories about the origin and diversity of life. The origin, as well as the diversity of life, in most human societies has often been ascribed to an act of special creation by an omnipotent higher power, often of anthropomorphic form. Many different explanations were in vogue in the early nineteenth century; the most widely accepted in many societies was the idea of "special creation"—that is, that each species of plant or animal had been specially created to fill some specific niche perfectly and that each species would thereafter remain unaltered and distinct. Charles Darwin was a "creationist" when he sailed on the *Beagle* in 1831. However, his observations of plants and animals inhabiting South America, and particularly the Galapagos Islands, together with his reading of Charles Lyell's then recently published *Principles of Geology* during the voyage, raised doubts in Darwin's mind about the validity of his beliefs. Shortly after Darwin's return to England in 1836, ornithologist John Gould assured Darwin that the mockingbirds Darwin had collected on three islands of the Galapagos, which had been isolated from the South American mainland and from each other for extended periods of time, were valid species. This gave Darwin the idea that new species might develop when a population had become isolated geographically and, hence, also isolated reproductively from its parental species. He deduced further that if colonists from a single South American species could diverge into three species on isolated islands of the Galapagos, then all of the fourteen species of mockingbirds on the South American mainland might also have branched off from a single ancestral mainland species as a result of ecological or other isolating mechanisms, and so also the species of related genera and increasingly higher taxonomic units as well. During this period Darwin also recalled his own and Charles Lyell's descriptions of geological discoveries that had made modern geology and paleontology possible. He wrote that these facts "seemed to throw some light on the origin of species ... that mystery of mysteries ... it occurred to me that it might be useful to accumulate and reflect on all sorts of facts that could have an effect on this question." This he continued to do for more than 30 years after the voyage of the *Beagle*, gradually identifying the four main premises of his theory of "descent with modification": (1) life is constantly changing, and all organisms are transformed over time, (2) all organisms evolved from a few common ancestors, and possibly all life traces back to a single organism into which life had been "breathed," (3) species multiply by splitting into descendent sibling species when isolated in geographically or ecologically divergent space, and (4) changes occur gradually over time and not by sudden production of entirely new types. It is not clear precisely when Darwin identified the fifth major premise of descent with modification, namely, that selection is the principal agent of change. In the *Origin of Species* he wrote:

> Preservation of favorable individual differences and variations, and the destruction of those which are injurious, I have called Natural Selection, or the Survival of the Fittest. At the commencement of my observations it seemed to me probable that a careful study of domesticated animals and cultivated plants would offer the best chance of making out this obscure problem. Nor have I been disappointed; in this and other perplexing cases, I have found that our knowledge, imperfect though it be, of variation under domestication, afforded the best and safest clue. I may venture to express my conviction of the high value of such studies, although they have been commonly neglected by naturalists.

Thus, Darwin acknowledged his debt to early plant and animal breeders in identifying selection as the chief directing agent in bringing about evolutionary change. Darwin recognized that breeders of plants and animals were aware that individuals within populations differ from each other and that offspring tend to resemble their parents. He was aware that breeders encourage the reproduction of individuals with desired characters, including characters that promote survival, and discourage reproduction of individuals with less desired attributes. He was also keenly aware that in natural selection no element of human choice is involved. Thus, survival within a natural ecosystem favors those individuals that are best adapted to the physical and ecological environment of that ecospace.

In 1859, after much self-debate and discussion with his many scientific, philosophical, and ecclesiastical friends, Darwin published *The Origin of Species by Means of Natural Selection or the Preservation of Favored Races in the Struggle for Life*.[1] The effect of Darwin's book on the study of biology and on humans' perception of their own place in the world was no less than revolutionary. Darwin was keenly aware of, and apprehensive of, the extraordinary intellectual and psychological adjustments his concepts would require of his readers and the outrage these concepts would be likely to engender. According to one of his biographers (Browne 1995), he recognized that any book he produced must be most carefully constructed and that he must be alert to the great demands he would be making on the sensitivities of his readers. He was fully aware that most people believed that a benign deity had created a perfect natural world in which each living thing had been specially and specifically "created" to fit perfectly into a predetermined place in nature. Yet his studies had convinced him that for natural selection to work, perpetual struggle, individual

---

[1]Alfred Russel Wallace, an English naturalist who studied variation for many years in the jungles of the Amazon and the Malay Archipelago, had arrived at the same conclusions as Darwin and had published his evidence and conclusions in a short paper in 1855 and in a second short paper mailed to Darwin and read before a meeting of the Linnean Society in London in July 1858. These two papers (reproduced in Brackman 1980), both of which predate the *Origin of Species* (1859), have led many scholars to identify Wallace as the codiscoverer of "descent with modification."

against individual, species against species, was essential—differential survival was the key to reproductive success. Thus, in Darwin's view, all qualities on which natural theologians and philosophers had based their ideas of perfect adaptedness had, in fact, arisen through competition—Malthusian (Darwinian) struggle. What most people saw as "deity-created" design, he saw as adaptations that were without meaning except as they allowed an individual plant or animal or an individual species to survive and reproduce. Darwin told some of his friends that he was not sure he could write such a book. However, the results of more than 30 years of intense and careful thought demanded expression, and by May 1856 he had convinced himself that he must proceed with the writing of a book, which he intended to call "Natural Selection." According to Browne (1995), he told a friend, "I am like Croesus overwhelmed by my riches in facts and I mean to make my book as perfect as it can be."

The *Origin of Species* (1859) provided a framework for the biological sciences that brought order to the overwhelmingly complex and confusing masses of observations on the morphology, physiology, ecology, and environmental interactions of plants with each other and with animals, as well as the progressions by which changes had occurred over long periods of time. It led to replacement of the concept of stasis in biology with a concept of recurring sequences of dynamic change in populations of varying individuals, resulting in branching processes that might culminate in ecotypes, varieties, species, and, sometimes, higher taxonomic categories. Considering the basic simplicity of the idea itself, it is not surprising that it was soon adopted by many biologists; for example, Thomas Henry Huxley is reported to have said shortly after reading the *Origin*, "How extremely stupid not to have thought of that." However, this was not generally the case—the book was indeed revolutionary. Darwin had thrust revolutionary ideas on the world, ideas that, in his words, led to "one long argument," one that has continued, but at slowly decreasing volume, up to the present. However, "one hundred thirty years of unsuccessful refutations have resulted in an immense strengthening of Darwinism . . . basic Darwinian principles are more firmly established than ever" (Mayr 1991), leading to a somewhat modified version of Darwinism often called Neo-Darwinian Evolution. Further, as stated in the first paragraph of this chapter, plant breeding has become increasingly efficient and focused as Darwinian and, more recently, Mendelian principles have become firmly established as the framework within which all evolutionary changes occur.

Evolution under Darwinian and Mendelian principles is essentially a two-phase process. The first phase is production of variation. Darwin was much aware of the ubiquity of variation in populations. He devoted Chapters 2 and 3, respectively, of his *Origin of Species* and much of a sequel volume, *The Descent of Man* (1872), to detailed descriptions of variation under conditions of domestication and in nature. But the origin of this obvious variation and the ways it was expressed puzzled Darwin. Indeed, to this day our understanding of the ways specific variants interact in "complex networks" to produce particular aspects of the phenotype is far from complete. Darwin also knew that variation is only partly heritable, that offspring have a mixture of the characters of both parents, and that expression of variations are often influenced by the environment. But that was all he needed to know to formulate an essentially valid theory of "descent with modification." An understanding of modern evolutionary genetics greatly sharpens our under-standing of the basic processes of evolution, but it is not a prerequisite to such understanding.

The second phase of evolution under Darwinian and Mendelian principles has to do with the reproductive potential and survival ability of variants—that is, the struggles for existence to which the rest of the *Origin of Species* (1859) as well as a sequel volume, *The Descent of Man* (1872), were directed. Reproductive potential can vary widely from individual to individual within species, and from species to species. Because populations usually do not differ greatly in numbers of individuals from generation to generation, it is apparent that something must act to control population growth. Many individuals are capable of producing dozens or even hundreds of offspring per generation, and such excess reproductive capacity might soon cause populations to exceed the carrying capacity of the environments they occupy. Darwin (1859) defined natural selection as the "preservation of favorable variants and the destruction of those that are injurious." Because the variations of individuals that do not survive to reproductive age are not passed on to the next generation, the characteristics of later generations will be more like those of the survivors than of the nonsurvivors. Thus, the qualities that determine survival in plant breeders' populations are, in large part, the same as the qualities that deter-mine survival in nature. However, additional factors enter the equation in breeders' populations; for example, breeders usually attempt to adjust the environment of their nurseries to correspond to the agricultural circumstances in which their selects are likely to be grown. Moreover, agricultural and consumer preferences may sometimes require that the breeder select variants with detrimental effects on reproductive and/or survival capacity. Nevertheless, understanding the concept of differential survival is plainly of key importance, because it makes clear that adaptations for survival in the broad sense are basic to the functioning of ecosys-tems, including agricultural ecosystems. The breeder strives to achieve a balance between adaptations that best satisfy multiple requirements, including those of both producer and consumer. Because both the environments in which agricultural plants are grown and the goals of plant breeding are critically important to the outcome of evolution under cultivation, we will consider these two aspects of differential survival in considerable detail in later chapters of this book.

# TWO

# *Origins of Agriculture*

There is general agreement that humanlike creatures had evolved in Africa by about 3 to 4 million years before the present (M.Y.B.P.), late in the Pliocene or early in the Pleistocene, and that much of the subsequent evolution of these hominids took place in Africa during the 3 million years or so of the Pleistocene (Johanson and Edgar 1996). One opinion is that about 1 million years ago an archaic form of *Homo* spread westward from northeastern Africa into Europe, as well as into southwestern Asia, and from there to the Far East. Archaeological evidence suggests that an early form of *Homo* had reached central China by 1.7 to 1.9 million years ago and that these early hominids may have been the ancestors of anatomically modern Asiatic forms of the genus *Homo*. Nearly all specialists on human origins agree that anatomically modern humans (genus *Homo*) originated relatively recently, perhaps no earlier than about 200,000 B.P. However, there is disagreement as to whether these anatomically modern humans evolved independently and more or less simultaneously in Africa, Asia, and Europe or originated in a relatively restricted area in Africa and spread from there to replace archaic populations elsewhere throughout the world. Those who favor a restricted origin (the out-of-Africa hypothesis) point to the occurrence of nearly modern types in Africa and its southwestern Asian periphery by about 100,000 B.P. Although the Africans and southwestern Asians of that period were anatomically modern, it is believed they were behaviorally primitive and that they did not develop ability to adapt to the environment through cultural advances until about 50,000 B.P. Recent archaeological evidence indicates that anatomically fully modern humans (*Homo sapiens*) were present in Central Siberia by about 55,000 B.P. and that these Siberians had a diverse tool kit similar to that of southwestern Asians of the time, including lightweight stone tools used for working bone and antlers into weapons and hunting gear, as well as for making ornaments (Morell 1995). This degree of cultural sophistication did not appear in Europe until a few thousand years later, which is consistent with the suggestions of some specialists that southwestern Asia was the center from which modern humans spread over the earth. Wherever they came from, these humans appear to have been as advanced anatomically as any who have come afterward. Their brains had high frontal cranial lobes, suggesting that they were capable of cognitive thought. They made animal-skin clothing and moccasins sewn with bone needles, and they constructed huts with sturdy wooden or bone frames, as well as snug tents. Some authorities believe that this people, equipped with ability to adapt to the environment through cultural sophistication, spread rapidly, reaching the Atlantic to the west and the Far East and Australia to the east by about 50,000 B.P.

A common impression is that the diets of Paleolithic humans were dominated by meat obtained through hunting or, more likely, by scavenging; this belief may have grown out of frequent references to animal bones found in prehistoric human sites and the visions they engender of early man as a successful hunter. It is possible that early archaeological evidence exaggerated the role of meat in the diet of most prehistoric humans; this may be because plant materials are much less likely to survive in prehistoric campsites than animal bones. Even recently abandoned sites of modern gatherers contain little in the way of plant remains, yet the diets of these peoples are known, from direct observation, to be primarily vegetarian. That plant remains, especially small seeds and small particles, could survive remained unlikely until the use of fire by humans became widespread; this is because charred plant remains, especially fine particles, withstand the ravages of time better than uncharred plant particles. However, small charred plant parts are not readily detected, and it is only in the last few decades that water flotation techniques have revolutionized the ability of archaeologists to find plant remains and that carbon isotope dating and other dating techniques have improved the ability to time-date such remains. Flotation techniques depend on placing materials from archaeological sites in water-filled containers, allowing inorganic sediments to settle, then skimming off floating plant remains. The results show that early humans were almost exclusively gatherers. The nutritional foundations on which they relied were primarily seeds of annual plants, but also tubers, rhizomes, and/or nuts from trees, supplemented by meat when a hunt was successful or relatively fresh carcasses were found. Analyses of the diets of present-day gatherer societies support this view. Most of the calories of present-day gatherer societies come from plants; only a few such societies, usually those located in resource-abundant riverine or maritime sites, depend heavily on fishing. Still fewer present-day gatherer societies, usually those located in cold, high-latitude regions, depend heavily on hunting. Neither meat nor dairy products became predominant items in human diets until sheep, goats, cattle, and pigs were domesticated in the Fertile Crescent about 8000 B.P. Moreover, the diets of most modern societies, nonindustrial as well as industrial, are based in large part on complex plant carbohydrates and proteins (Table 2-1). Many specialists predict that this dependence on plants is likely to become even greater as human population numbers rapidly increase in the coming decades.

Modern gatherers harvest seeds of numerous different plants, especially large-seeded grasses and legumes, as well as tubers, rhizomes, fruits, and other parts of many plants. They also occasionally "manage" some plants with fire and protect certain plants while destroying others. This finding has led to speculation that earlier gatherers might have affected the production and survival of different plant species and genotypes within species differentially, thereby modifying the genetic composition of some of the species on which they depended. Although gatherers probably depended on plants as their dietary mainstay throughout the 750,000 or so years of the Paleolithic Age, few archeological sites of the period have been identified and excavated, and few plant remains of the period have been identified.

Consequently, the interrelationships between humans and the plant world during the long Paleolithic Age remain uncertain. The general consensus is, however, that gatherers had little impact on the evolution of plants, even the plants they gathered. In fact, archaeological records of plants are limited, in large part, to the last 15,000 years, and it is to the Mesolithic and Neolithic Ages that we must turn for more reliable information concerning our next topic, the origins of agriculture, leading to the transformation of some foraging human societies into agriculture-based societies.

## ORIGINS OF AGRICULTURE

Evidence bearing on the origins of agriculture comes from many disciplines, including anthropology, archaeology, genetics, ecology, molecular biology, sociology, and taxonomy. Recent reviews of the evidence indicate that early agricultural economies developed independently during the late Mesolithic and early Neolithic Ages (~15,000 to ~10,000 B.P.) in many different places in the world (Harlan 1992; Cowan and Watson 1992). Embryonic agricultural economies usually featured the gathering of seeds, tubers, and other plant parts from several up to many different species. Moreover, gatherers may sometimes have imposed elementary husbandry practices on the plant populations they intended to harvest such as sowing seeds on sites that had been deliberately fired to lessen competition, seeding or transplanting on the moist edges of retreating ponds (*playa*, or "beach," agriculture), or deliberate irrigation by diverting streams onto relatively fertile sites. Apparently, once human societies became committed to particular suites of wild plants for sustenance, some degree of residential stability (sedentarianism) often followed. Yet residential stability did not always lead to agriculture. Many gatherer societies were sedentary but did not become agricultural. For example, those of the Pacific coastal areas of North America were sedentary but remained nonagricultural, perhaps because natural resources, both plant and animal, especially fish, were abundant and dependable. Although there were many differences from place to place in the development of agriculture, the evidence suggests numerous commonalities in agricultural origins from continent to continent.

Cowan and Watson (1992) have postulated two major settings, one in which locally domesticated plants were gradually added to a foraging-based economy. This in situ process was one of coevolution of agricultural and cultural systems without the introduction of primary crop plants from outside. These authors regard the Near East and Mesoamerica as particularly good examples of this pattern. They also envisage two sorts of situation in which agriculture-based economies developed as "secondary" phenomena. In one case a suite of domesticated plants was introduced from outside into an economy previously dominated by foraging. Whether single crop by single crop, or by wholesale introduction of many domestic species, agriculture slowly replaced foraging in the recipient regions, as appears to be the case in Europe and the North American Southwest. In the second case, an exotic domesticate (corn), ultimately highly successful, was introduced into an economy already dependent, at least in part, on agricultural plants. Once such an introduction had taken place, the new plant frequently, sooner or later, became dominant in the economy. As a result of extensive and intensive water flotation and carbon isotope studies carried out in the period 1980–1990 in eastern North America, this area now provides one of the most detailed records presently available of the origins of agriculture. The eastern North American transition from forager to farmer economies involved (1) the domestication of at least four indigenous seed plants (sunflower, chenopods, marshelder, cucurbits) by 2500 B.P., (2) the emergence of food-production economies based on locally domesticated crop plants by 2000 B.P., (3) a rapid and broad-scale shift to a corn-oriented agriculture in the three centuries 1200 to 900 B.P., followed by (4) the addition of many exotic domesticates in the five centuries after Columbus (500 B.P.). The chronologies of agricultural origins are well understood in many other areas of the world (reviewed in various of the papers in Cowan and Watson [1992]). However, in no case is it clear why, in a period starting at about 15,000 to 10,000 B.P., so many gatherer societies throughout the world turned away from foraging as a way of life to adopt agriculture. Numerous authors have cited evidence that the diets of gathering peoples were good, periods of starvation were rare, their health was good, and their life-styles were varied and satisfying. Nearly all students of agricultural origins have concluded that the reasons for the widespread shift from gathering to agriculture that occurred in the very late Pleistocene or early Holocene are elusive and likely to remain so.

## THE GEOGRAPHY AND ECOLOGY OF AGRICULTURAL ORIGINS AND PLANT DOMESTICATION

No discussion of these topics would be complete without mention of Charles Darwin, Alphonse de Candolle, and Nikolai Vavilov. Darwin (1868) contended that domesticates differ little from their wild ancestors in amenability to agricultural circumstances and that the main changes resulting from selection during the process of domestication were increases in the size and change in the form of organs most valued by humans. Similarly, de Candolle (1866) wrote that domesticates often differ much more from one another than from their wild ancestors. Vavilov (1926) originally believed that the geographical region in which genetic diversity was greatest corresponded to the geographical region of origin, especially if wild races of the relevant species were also present in that region. He defined eight centers of origin by drawing lines around areas in which human civilizations had arisen and in which agriculture had long been practiced. Later he concluded that his concept of centers of origin often did not work well, and he developed a system of ecological groups based on such traits as response to length of photoperiod, response to temperature, reactions to diseases, and specific adaptedness within particular environments.

Substantive modifications in the concept of centers of origin have evolved since the time of Vavilov. As centers of origin have been analyzed one by one, it has become apparent that most crops did not originate in a center of origin such as proposed by Vavilov: patterns are much more complex and diverse than Vavilov had visualized. There does seem to have been a clear center in the Near East (Fertile Crescent) in that several of our major crop plants and animals were domesticated in that relatively small region, the domesticates diffused outward from the center into adjacent ecologically similar regions, and many later spread throughout the world. However, Vavilov's other centers have become increasingly diffuse as more information has become available. Harlan (1992) suggested that the time may have come "to abandon the concept of centers altogether" and refer instead to "ecological regions rather than to centers."

There is now substantial agreement that four complementary and overlapping ecological-evolutionary-genetic processes were responsible for present-day eco-geographical patterns of genetic variability:

1. Specific late Pliocene or early Pleistocene ecological settings guided natural selection in directions that preadapted or predisposed a few wild species with properties that later led to success in agriculture.
2. Some reproductive systems, especially selfing and vegetative reproduction, provided opportunities for rapid generation of novel superior genotypes in cultivation, at the same time providing that the best genotypes present in cultivated populations would be able to reproduce themselves unchanged and not be lost quickly owing to segregation resulting from hybridizations with inferior genotypes; thus, selfing and vegetatively reproducing populations adjusted rapidly to conditions of agriculture.
3. The superior domesticates soon spread beyond their cradles of domestication into geographic areas with different environments.
4. Increases in adaptedness within specific environments led to still greater genetic differentiation among populations of domesticates.

The first step probably took place in the late Pliocene and the last several thousands of years of the Pleistocene, whereas the second step probably took place in the Mesolithic or the Neolithic Age coincident with the dawn of civilizations. The third step probably began soon thereafter and accelerated greatly with the episodes of discovery and colonization of the last five centuries. The fourth step, modern plant breeding (the main topic of this book) did not start until late in the nineteenth century, and much of the gain from this step has come in the twentieth century. An obvious starting point in identifying plants with characteristics conducive to wide success in agriculture is to identify species that have been successful on the global agricultural scene and to compare the circumstances of their origin with those of species that have been unsuccessful or less successful. Although there is no entirely satisfactory way to compare the global importance of different cultivated plants with one another, or with animal products, Food Production Yearbooks of the United Nations give estimates of the worldwide production of edible dry matter and protein for widely grown crops and major animal food products. Fortunately, these estimates nearly always lead to conclusions that are in accord with long-term subjective assessments of the value to humans of various food sources; thus, they give numerical substance to value judgments made over the centuries by farmers and consumers.

## CROPS THAT FEED THE WORLD

Table 2-1 gives global production estimates of edible dry matter and protein for the leading food crops and for major animal products, all expressed as percentages of total world production. The table includes numerical values for only those individual groups (or species groups) that contribute about 1% or more annually to the total world production. Under this criterion only eight crop commodity groups (cereal grasses, cereal legumes, oil seed, root, tuber and starchy foods, sugar crops, vegetables, fruit crops) and three animal commodity groups (milk, cheese, and eggs; meat; fish) qualify in both the dry edible matter and protein categories. Among individual plant species all of the eight cereal grasses listed in Table 2-1 (wheat, corn, rice, barley, sorghum, oats, rye, millets) qualify in both the dry edible matter and protein categories. However, in other commodity categories only six individual crops (soybean, peanut, cassava, sweet potato, cane sugar, sugar beet) qualify in the edible dry matter category and only four individual crops (soybean, peanut, bean, potato) qualify as producers of protein. Estimates for vegetables and fruits are less certain, but these two commodity groups jointly probably contribute only about 3.5% of the total edible matter and only about 4% of the protein to total world food production; however, the overall nutritional value of vegetables and fruits probably exceeds their caloric contributions. At the global level all animal products, including fish, contribute about 6.5% of the edible dry matter but a much higher percentage (about 17%) of the protein. Globally, forages (largely grasses and legumes), along with many browse species, are estimated to contribute more than 75% of the energy provided in the diets of livestock, whereas cereals and cereal legumes are estimated to contribute about 20%. Perhaps 60% of total cereal production (especially barley) in developed countries is now used for animal feed, but only about 20% in developing countries.

Many species, in addition to the major plant species listed in Table 2-1, have been cultivated and/or domesticated; however, it is obvious that most of the food of humans comes from a very small number of crop plants. The three major cereals (wheat, corn, rice) contribute about 54% of the total, and if barley is added, this value increases to more than 60%. This does not disparage either the combined contributions or the dietary importance of the lesser crops, or the contributions of the dozens (perhaps even hundreds) of species that are still less frequently consumed by humans or livestock. Nevertheless, the overall impact of these numbers is overwhelming. The worldwide food supplies of humans depend in very large part on the 30 or so crop species shown in Table 2-1. Among these only a few—the major cereals (wheat, rice, corn, barley); supplemented by one or more legumes (soybean, peanut, pea, phaseolus beans); a few tuber, root, and starchy crops (potato, sweet potato, cassava, bananas, and plantains); and sugar crops (cane and beet)—provide the major nutritional needs of nearly all major human civilizations. Furthermore, the current trend is for the major crops to increase in importance and lesser crops to dwindle. This brings us back to the question of the evolutionary processes that predisposed some plants to suitability for cultivation and/or domestication. Cultivation involves activities of humans, such as tilling the soils, seeding, weeding, fertilizing, and watering, that alter ecological conditions, thereby setting the stage for selection favoring types adapted to the ecological conditions engendered by

TABLE 2-1  World production of total edible dry matter and protein by leading crops and animals, expressed as percentages of world totals[*]

| COMMODITY GROUP | TOTAL EDIBLE DRY MATTER | PROTEIN | REPRODUCTIVE SYSTEM IN NATURE/ CULTIVATION | ECOLOGICAL ORIGIN IN NATURE |
|---|---|---|---|---|
| *Cereals* | | | | |
| Wheat | 20 | 19 | S | M |
| Corn (maize) | 18 | 13 | C/C | S |
| Rice | 17 | 11 | S | S |
| Barley | 7 | 5 | S | M |
| Sorghum | 3 | 2 | S/C | S |
| Oats | 2 | 2 | S | M |
| Rye | 1 | 2 | S | M |
| Millets | 1 | 1 | S | S |
| Subtotal | 69 | 55 | | |
| *Legumes* | | | | |
| Soybean | 3 | 9–10 | S | S |
| Bean | <1 | 1 | S | S |
| Pea | <1 | 1 | S | M |
| Peanut | ≤1 | 1 | S | S |
| Subtotal | 6 | 13 | | |
| *Vegetable Oil Seed* | | | | |
| Rapeseed | 1 | 1 | C | M |
| Sunflower | <1 | 1 | C/C | S |
| Cotton | <1 | <1 | S | S |
| Coconut | ≤1 | ≤1 | V | T |
| Subtotal | 3 | 2 | | |
| *Tubers, Roots, Starchy Foods* | | | | |
| Potato | 3 | 2 | V | TH |
| Cassava | 2 | 1 | V | S |
| Sweet Potatoes | 2 | 1 | V | S |
| Yams | | | | |
| Banana, Plantain | ≤1 | ≤1 | V | S |
| Subtotal | 8 | 5 | | |
| *Sugar Crops* | | | | |
| Cane | 3 | 0 | V | TF |
| Beet | 1 | 0 | C | M |
| Subtotal | 4 | 0 | | |
| *Vegetables* | | | | |
| Tomato, Cabbage | 1 | <1 | S,C,C/C | S |
| Onion | — | — | C | M |
| Subtotal | 2 | 1 | | *(continued)* |

TABLE 2-1 Continued

| COMMODITY GROUP | TOTAL EDIBLE DRY MATTER | PROTEIN | REPRODUCTIVE SYSTEM IN NATURE/ CULTIVATION | ECOLOGICAL ORIGIN IN NATURE |
|---|---|---|---|---|
| *Fruits* | | | | |
| (Grape, Apple, | | | V | W,W,T,T |
| Coconut, | | | | TF |
| Oranges) | — | — | | |
| Subtotal | <1 | <1 | | |
| *Animal Products* | | | | |
| Milk, Cheese | 3 | 6 | C | M |
| Eggs | | | | |
| Meats | 3 | 7 | C | M |
| Fish | 1 | 4 | C | |
| Subtotal | 7 | 17 | | |
| *Other* | 2 | 7 | | |

*Based on data from *FAO Production Yearbook*, 1987, and *FAO State of Food and Agriculture*, 1987.

**Reproductive System:** S = selfing; C/C = originally cross-fertilized but recently increasingly grown as $F_1$ hybrids; S/C = originally largely selfed but recently increasingly grown as $F_1$ hybrids; V = vegetatively reproduced.

**Ecological Origin:** M = Mediterranean; S = savanna; T = tropical lowlands or tropical coastal; TH = tropical highlands; TF = tropical forests; W = temperate woodlands.

cultivation. Moreover, humans select types they value and thus, by direct artificial selection, may produce or accelerate the development of genetically modified populations (cultivars). The distinction between cultivars and domesticates is ill defined. Some consider plants to be domesticated when they have been sufficiently altered that they can easily be distinguished from their wild progenitors, whereas others restrict the term *domestication* to plants that have become so altered that they can no longer survive without human intervention. The threshold of complete dependency on humans is rarely passed until many generations after the initiation of cultivation, and it is probably not possible to pinpoint the precise time when any domesticate passes the threshold to cultivar. Henceforth we will be little concerned with this difficult distinction but will apply the term *cultivar* to any form that can readily be distinguished from its wild ancestor.

Although there is justification for believing that domestication may have started earlier in Africa, Southeastern Asia, and/or Central America than in Southwestern Asia, discussions of plant domestication usually begin with those parts of modern Jordan, Israel, Lebanon, Syria, Turkey, Iraq, and Iran now often known collectively as the Levant or the Fertile Crescent. There are many reasons for this. One is that the oldest civilizations about which we have much information are those that developed in the Fertile Crescent. Prior to about 11,000 B.P. groups of mobile foragers lived by gathering and hunting in this area, but by 7000 B.P. much of the region was inhabited by settled villagers who relied heavily on farming and raising livestock for their livelihood. The transition from foragers to villagers in the Fertile Crescent is documented by many artifacts that identify innovations associated with the beginnings of the cultivation of plants (e.g., tools used in tilling the soil and harvesting, as well as pottery used to store grains). Another reason relates to the Mediterranean climate of the Fertile Crescent. The mild, wet winters and hot, dry summers favor the evolution of annuals with large seeds that survive the summers in dormancy, then germinate with the first wetting rains in the fall, utilizing large stores of seed carbohydrates and proteins to outgrow less vigorous seedlings of competing plants, and ultimately produce large yields of nutritious and readily harvested grains by late spring. Moreover, it is often dry at flowering time in the spring in Mediterranean climates. This favors the evolution and survival of plants with cleistogamous flowers, flowers that remain closed during much of the pollination period. Cleistogamy inhibits desiccation of anthers and stigmas, thus promoting full seed-set by self-pollination; yet brief periods when flowers are open allow occasional hybridizations to occur, especially between close neighbors in the same population. Genetic segregations in the first few generations following such hybridizations quickly generate novel true-breeding (homozygous) genotypes, and at the same time predominant selfing ensures that such superior genotypes will not quickly be lost as a result of segregation when hybridizations occur with less well adapted neighbors. It seems likely that this continuing and self-sustaining process of occasional crossing, usually followed by repeated self-fertilizations, played an important role in the evolution of the high adaptedness of predominantly self-pollinating plants to the Mediterranean climate of the Fertile Crescent, thus predisposing so many self-pollinators to domestication (Table 2-1). This feature of predominant selfing was also almost certainly good for early farmers, who no doubt saved individual variants with features they valued (e.g., nonshattering inflorescences that have poor survival ability in the wild but retard seed loss, thus improving agricultural harvests). Such predominant selfing may have been an important factor underlying the ability of self-pollinating plants to adjust so rapidly to conditions of agriculture in the Fertile Crescent, as well as to new environments when humans later dispersed them throughout the world.

The earliest evidence of morphologically modified plants in the Fertile Crescent appeared in the archaeological record at about 11,000 B.P. Domesticated barley (derived from wild *Hordeum vulgare*, ssp. *spontaneum*) has been reported from the earliest archaeological sites, although there is some question whether these early barleys were morphologically distinguishable from wild types. The most primitive domesticated einkorn wheat (*Triticum monococcum*, N = 7 chromosomes, derived from wild *T. beoticum*) and the most primitive emmer wheat (*T. dicoccum*, N = 14 chromosomes, derived from a wild tetraploid wheat, *T. dicocoides*) first appeared in archaeological sites dated about 10,000 B.P. Very rapid evolution toward morphologically domesticated types may explain why intermediate forms have been encountered only rarely in the archaeological records. Bread wheat (hexaploid *T. aestivum*, N = 21 chromosomes) was present in some archaeological sites by about 8,000 B.P.; it apparently arose in cultivation from segregants from hybrids between tetraploid cultivated emmer (N = 14 chromosomes) and a weedy wild goat grass,

*Aegilops squarrosa* (= *T. tauschii*) (N = 7 chromosomes). Wild rye (*Secale cereale*, N = 7 chromosomes) also occurs in the Fertile Crescent, but rye (an outbreeder) apparently became established as a crop relatively late. Most of the calories consumed by the earliest farmers come from these high-carbohydrate cereals, which were the most useful of the cereals of the Fertile Crescent because of their abundance, large seed size, high seed yields, and relative ease of harvest in both wild and cultivated stands.

Carbohydrates and proteins are both critical items in human diets. The cereal grasses of the Fertile Crescent are an excellent source of carbohydrates and are also fairly high in protein. But unlike animal protein, cereal protein is unbalanced; it is particularly low in lysine, one of the essential amino acids required by animals in the synthesis of protein. However, during the beginnings of farming in the Fertile Crescent, the cereal grasses were soon complemented by two other types of food with higher protein content: (1) cereal legumes, especially lentils, vetches, and peas, which are high in protein and, in addition, are high in lysine and (2) domestic animals (sheep, goats, cattle, pigs), which provide excellent amino acid balance and also produce leather and/or wool. Still another largely self-pollinating plant, flax (*Linum usitatissimum*, derived from wild *Linum bienne*), with its fat-rich seeds, completed the dietary trinity (carbohydrates, protein, fat); flax also provided a useful textile fiber. In short, the early farmers of the Fertile Crescent developed a suite of domesticated plants and animals that satisfied all of the basic needs of humans: carbohydrates, protein, fat, and textiles, as well as animal products used for food and for other purposes. As these cultigens and/or domesticates radiated from the Fertile Crescent westward into the Mediterranean Basin and North Africa, southward to Ethiopia, and eastward to Afghanistan and Pakistan, they were apparently quick to adjust genetically to each new set of environmental circumstances. These adjustments continued as they radiated farther outward to include virtually all of the two Temperate Zones (~ 23° 30' to ~ 66° 30' north and south of the equator), thereby taking a conspicuous place among the crops on Table 2-1.

As emphasized by Harlan (1992), many of the remaining plants listed on Table 2-1 evolved and were domesticated in savanna settings. Savannas are tropical or subtropical grasslands (usually interrupted with scattered patches of shrubs or trees) that occur in areas in which a warm, rainy season is interrupted by a long, cool-season dry spell. The ecological setting of a long dry spell not only favors large-seeded annuals, as discussed earlier, but also favors plants with rhizomes, tubers, stolons, and bulbs that go dormant in the dry season, causing such plants to behave much like annuals. With the onset of warm-season rains, the rhizomes, tubers, stolons, and bulbs break dormancy and utilize their stored food reserves to produce rapid growth of roots below ground and shoots above ground. By the time the rainy season ends, nearly all of the food produced in the aboveground parts has been translocated downward, the aboveground parts die back, and the storage organs go dormant, remaining so until the start of the next rainy season, when buds of the dormant organs break dormancy to initiate the next generation. An important feature of this reproductive system is that all vegetatively produced propagules from each parental individual are identical genetically and breed true for all characters. Hence, once a superior genotype appears, whether in a wild or in a cultivated population, its superiority gives it an advantage in survival. On the other hand, most vegetatively propagated plants are cross-pollinated and highly heterozygous, and if able or allowed to reproduce sexually, produce many novel genotypes. Vegetative (clonal) reproduction thus provides for the production of large numbers of superior genotypes through segregation and recombination, but also allows for long-term preservation of superior types through vegetative reproduction once such types appear. Consequently, like selfing, the system is favorable for rapid and stable evolutionary adjustments when such plants encounter new environments. The main savanna plants with the aforementioned characteristics on the list of common food crops (Table 2-1) are cassava (*Manihot*) and sweet potato (*Ipomoea*) from the warm lowlands of South America and yams (*Dioscorea*) from various parts of both the New and Old Worlds.

Large-seeded annuals are favored in tropical and subtropical savannas as well as in Mediterranean climates, and many such annuals appear on the list of 30 food plants (Table 2-1)—for example, rice, corn, sorghum, millet, beans, peanut, and cotton (which is perennial in nature but annual in domestication). Wild rice (*Oryza sativa*) from Southeast Asia grows in standing water, which collects in depressions during the rainy season, and survives the dry season in the dormant seed stage. Most wild savanna plants other than rice evolved under somewhat drier circumstances, especially the savanna plants of Central Africa, Central America, and South America. Pearl millet (*Pennisetum glaucum*) from the dry southern fringes of the Sahara desert is one of the most drought resistant of all crop plants. Sorghum (*Sorghum bicolor*) is less drought resistant; it is from a broad, less xeric belt stretching east to west across Central Africa south of the pearl millet belt but north of the moist rice area in which African rice (*Oryza glaberrima*) originated. Central America and South America are areas of great environmental and ecological diversity. In very general terms, three climatic regions can be distinguished in Central and South America: (1) zones of subtropical semiarid savannas characterized by distinct winter and summer temperature seasons, (2) tropical lowlands, and (3) cool tropical highlands. The only major cereal from the Americas is corn, which was probably first domesticated in the midelevation, semiarid savannas of Southwestern Mexico. Teosinte (*Zea mays*, ssp. *mexicana*, an outcrosser), the putative wild ancestor of corn, produces paltry yields of small, hard seeds even under the most favorable conditions, and it is difficult to deduce what its original appeal to humans may have been, especially as a cereal. It has been suggested that its original appeal may have been as a vegetable; young inflorescences are palatable, and immature corn inflorescences are presently sometimes used as a vegetable. Nevertheless, largely in the twentieth century, corn was subjected to human-directed genetic changes that ultimately converted it into one of the three major cereal grain crops of the world. Two major annual cereal legumes, lima beans (*Phaseolus lunatus*) and common beans (*P. vulgaris*), were among the earliest plants to be domesticated. The wild ancestor of the small-seeded lima bean is found in semiarid brushlands in Central America and southern Mexico, whereas the wild ancestor of the large-seeded lima bean occurs on the western slopes of the Andes. The wild ancestor of the common bean (*Phaseolus vulgaris*) ranges from Mexico down the eastern slopes of the Andes to northern Argentina. Several species of truly wild tomatoes (*Lycopersicon escalentum*) occur on the western coast of South America. However, the tomato apparently was not domesticated there but, rather, in Central America or southern Mexico, where weedy, small-fruited tomatoes are common in disturbed areas. The inbreeding wild peanut (*Arachis hypogaea*) occurs in midelevation savannas in northern Argentina and adjacent Bolivia. Upland cotton (*Gossypium hirsutum*) was domesticated in southern Mexico. Wild forms and early domesticates were perennial, but annual types were developed as cotton became a major industrial fiber and a major oil crop in more temperate zones. The cool highlands of South America contributed one of the most important of all food crops on the world scene, the white potato (*Solanum tuberosum*).

Tropical forests (sugar cane [*Saccharum officinarum*], bananas [*Musa*], orange [*Citrus*], mango [*Mangifera*], *and coconut* [*Cocos*]), as well as temperate woodlands (apple [*Malus*], peach [*Prunus*], sunflower [*Compositae*], *grape* [*Vitus*]) also contributed many domesticates that are important today throughout the world. Most of the wild ancestors of these crops are, however, adapted to forest margins, where they receive more sunlight than available under tightly closed forest canopies.

# THREE

# *Evolution During Domestication*

Despite vigorous and cogent defenses of Darwinian evolution by August Weismann (Germany) late in the nineteenth century, opposition to Darwin's theory of "descent with modification" continued into the twentieth century. Weismann's arguments were necessarily speculative, primarily owing to lack of relevant experimental evidence concerning heredity at the time; consequently, his efforts had little immediate effect. Weismann did, however, provide intellectual preparation for quick acceptance of three decisive sets of experimental results that appeared in the first and third decades of the twentieth century. First came the rediscovery of Gregor Mendel's experimental results, published in 1865 in an obscure journal, and unknown until brought to the attention of the scientific community principally by Correns (1900). With the emergence of Mendel's laws it became clear that the genetic material is made up of alternative variants of separate units (genes)—that is, that it is constant (none of the changes that accrue in the phenotype of an organism during its lifetime can be passed on to its offspring). Thus there is no blending inheritance and no inheritance of acquired characters. Another major set of experimental results (Johannsen 1903) established that the relationships between genotype and phenotype for continuously varying traits are compatible with Mendelism. A third set, by E. M. East (1916), established beyond doubt that the inheritance of corolla length, a continuously varying character in *Nicotiana longiflora,* is not blending but is due to segregation at several genetic loci, each with individually small but cumulatively large effects on the phenotype. However, it was not until near the middle of the twentieth century that a broad and coherent picture of evolutionary change emerged.

From the early 1900s to the mid-1940s most scientists interested in evolution were specialists in taxonomy, genetics, paleontology, ecology, physiology, or some aspect of agriculture such as plant or animal breeding. These specialists differed widely in biological background, in the hierarchical levels of biological organization with which they dealt, and also in their interests and goals. As examples, the primary concern of geneticists was with intrapopulational variation at the level of the gene, that of taxonomists was with morphological differences among species, that of paleontologists with the origin of higher taxonomic categories and that of breeders with improving agricultural performance. The differences were great enough to preclude rational and meaningful communication among groups. Yet within a 20- to 30-year period a broad consensus had been reached in the form of a second Darwinian revolution termed the evolutionary synthesis by Huxley in his treatise *Evolution: The Modern Synthesis* (1942). As discussed by Mayr (1991), the changes that occurred during this period did not stem from discovery of novel concepts. Rather they came in part from mutual education leading to rejection of erroneous theories concerning their own disciplines that had been held by previously feuding specialists and in part through the combining of germane parts of several disciplines into a cogent theory. Thus, as one example, geneticists, particularly genetic reductionists who had believed that each gene plays a largely independent role in fitness and thus that evolution is little more than change in allelic frequencies in populations, did not accept the Darwinian postulate that selection targets entire individuals. It also came to be recognized that neo-Darwinian evolution is compatible with and almost *certainly responsible for macroevolution (evolution above the species level)*.

The main features of the "evolutionary synthesis" can be summarized as follows: (1) genetic constancy was accepted and "blending inheritance" was appropriately laid to rest; (2) spontaneous mutations at individual genetic loci were recognized as the primal source of all genetic variability; however, segregation and recombination among individual variants that had accrued over time came to be seen as the immediate and main source of genetic variability available for selection; (3) individuals within populations rather than individual genes were recognized as the foremost targets of selection; (4) differing reproductive rates (fecundity) and different survival rates among genotypes were recognized as the main directing forces of "descent with modification". This overall picture has been amplified in the postsynthesis period but the principal features have not changed in fundamental ways.

The leading features of the evolutionary synthesis as it applies to wild ancestors and cultivated plants alike are as follows. All species are differentiated into genetically distinct populations as a result of natural selection operating on genetic variability in the various ecogeographical locations in which any species has gained a foothold. Thus, at any given time, a species will almost certainly be made up of many (often a very great many) intraspecific groups (e.g., subspecies, races, ecotypes, populations, strains, cultivars), between which varying amounts of genetic exchange are possible. Occasionally, if reproductive isolation is complete or virtually complete for sufficiently long periods of time, such groups may diverge to the point where they are no longer able to exchange genetic materials and may separate into distinct species. The essential features of neo-Darwinism thus emerged as population processes and as genetic processes in which adaptedness in each local environment is attained by successive trial-and-error allelic substitutions at many genetic loci, leading to local differentiation in genotypic frequencies and, occasionally, to speciation. In the next sections of this chapter we will see that cultivated plants obey precisely the same evolutionary rules as their wild ancestors—with the difference that human alterations, particularly in rates of genetic exchange between populations and in selective pressures, often alter the direction and/or the rates of evolutionary change once plants are taken into cultivation. However, even though human interference often leads to dramatic morphological and/or physiological changes in cultivation, there appear to have been no clear-cut cases of cultivated plants in which divergence has progressed to the point of speciation.

## SELECTION UNDER CULTIVATION AND DOMESTICATION

As discussed in Chapter 2, a striking series of changes took place in the human societies of the Near East in the period from about 11,000 B.P. to about 7000 B.P. Prior to about 11,000 B.P., groups of mobile foragers lived by gathering and hunting in this area. However, by 7000 B.P. much of the Fertile Crescent was inhabited by settled villagers who relied primarily on farming for their livelihood. Cultivation and domestication of plants and animals played a major role in this process and, ultimately, also had dramatic cultural implications for humans as well. Although the archaeological record of early plant cultivation and domestication is incomplete, it is nevertheless clear that the humans of the region were already heavily dependent on cultivated cereals for a major part of their diet before most cultigens had reached the stage of full domestication.

In understanding the genetic changes involved in domestication, it would be particularly helpful to determine the relative roles played by natural selection versus selection by humans. Natural selection under conditions of cultivation is expressed as inherent differences in reproductive capacity and/or in survival abilities whenever genetically heterogeneous populations of plants are cultivated in particular places (environments) at particular times. Human-directed selection occurs when farmers (or plant breeders) make conscious choices to keep the progeny of some plants while shunning the progeny of others. Evidence is presented later in this chapter (and in subsequent chapters) that natural selection in itself frequently leads to improved adaptedness, expressed as greater reproductive capacity and/or improved survival ability in the environmental conditions in which successive cultivated generations are grown—that is, that farming in itself can play an important role in adapting wild plants to cultivation. At the same time, whenever farmers (or plant breeders) choose plants with characteristics they value, plants with these characteristics are likely to increase in frequency in subsequent generations. Although the choices made by humans usually improve value to consumers, those choices may sometimes be inimical to superior reproductive capacity and/or survival ability. Thus, compromises may be necessary when usefulness (as perceived by humans) and improved adaptedness (greater reproductive capacity and/or survival ability under conditions of cultivation) are in conflict. We now turn to some examples in which natural and human (artificial) selection play opposing roles.

Examples from which it can be inferred that human selection played a predominant, and perhaps even sometimes an exclusive role, include selection against toxic or bitter constituents or against objectionable morphological traits such as thorns or irritating bristlelike outgrowths on stems, leaves, or other organs. Such traits are presumably favored in nature because they protect against predation. In cultivation, however, more palatable or nonprickly plants are, for obvious reasons, favored by humans even though such features may sometimes reduce yielding ability. For example, although rough-awned (wild type) barley isogenic lines usually yield more than their smooth-awned cultivated counterparts, smooth-awned cultivars are usually preferred by farmers because they are less irritating to workers during threshing. Considering the usual reproductive inferiority of the smooth-awned types, it is difficult to imagine that natural selection played much of a role, if any, in the emergence of smooth-awned cultivars. Similarly, it is difficult to imagine that natural selection played much of a role in reducing such traits as the toughness of peels, the numbers of hard seeds in fruits, or the hardness of shells of nuts, whether these traits are favorable or only neutral in respect to reproductive capacity and/or survival. Again, human preferences are obvious and human selection must have been the major and, perhaps in some cases even the exclusive, force involved in the development and adoption of types that are valued by humans but are lower yielding in cultivation.

Both human and natural selection appear to have been involved with many other characters, but at intensities that have varied widely from character to character and have probably shifted in importance in early versus later stages in crop evolution. Among the adaptations that separate wild from cultivated cereals, degree of seed retention in inflorescence is perhaps the most conspicuous. Several different mechanisms enhance seed retention in cereals that have unbranched culms (e.g., barley, wheat). In wild barley the rachis at maturity disarticulates at each node into individual units, each composed of a single seed and two sterile lateral spikelets; thus the seeds, at maturity, fall to the ground and nearly always escape harvest by humans. However, the rachises of some wild barley plants shatter less easily, and it seems likely that when humans began to harvest wild barley for planting, such plants would have been favored for their partial, but nevertheless helpful, seed-retention abilities. Repeated over generations in cultivation, this preference presumably supplemented natural selection in the development of cultigens with increasingly tougher rachises. Indeed, some of the cereal seeds found in the earliest archaeological sites of the Near East have had short internodal fragments attached to them; apparently, disarticulation at the nodes was incomplete in some of the earliest cultivated barleys, and the rachises of these barleys broke in internodal regions when harvested spikes were threshed. In present-day wild barleys a closely linked pair of complementary loci jointly give rise to a fragile (brittle) rachis that disarticulates at all rachis nodes when dominant alleles $Bt_1$ and $Bt_2$ are present, whereas modern cultivated barleys have nearly completely nonshattering rachises when alternative recessive alleles $bt_1$ and $bt_2$ are both present. $F_2$ hybrids between wild and cultivated barleys segregate in an approximately 3:1 ratio (three shattering: one tough rachis) as expected, with two fully dominant alleles at two very tightly linked loci. However, there are many intergrades of seed retention within the tough rachis class of different $F_2$ barley hybrids and even more intergrades in large $F_3$ and subsequent filial generations. This is because the strength of attachment of seeds to the nonshattering rachis in modern barley cultivars depends on many genetic loci and differs in strength from variety to variety; for example, some culitvars adapted to moist climates are prone to drop seeds from still-intact rachises when such cultivars are grown in dry climates, whereas some dry-climate cultivars are very difficult to thresh when grown under humid conditions. The preceding observations suggest that a very strong multilocus system of modifiers of loci $Bt_1$ and $Bt_2$ developed that converted rachis shattering into a very sharp either/or system (brittle vs. nonbrittle or shattering vs. nonshattering rachis), and that an alternative system of seed retention based on several genetic loci that jointly give rise to different levels of strength of attachment of seeds to nonshattering rachises developed more or less simultaneously. This has likewise been the case in most other major cereal species: in many species shattering versus nonshattering of the rachis is under relatively simple one-locus or two-locus control, as in barley. However, supplementary means of seed retention (or loss) differ from species to species; for example, in wheat the glumes are reduced to different extents and they enclose the kernels more tightly or loosely to give different levels of free-threshing in different genotypes. In other species whose wild relatives have lateral seed-bearing branches (e.g., corn, sorghum, sunflowers), there has been a strong tendency toward coalescence of many small infloresences into a single large inflorescence; this improves seed retention as well as ease of harvesting in cultivation. Dozens of additional examples could be given of morphological changes that have occurred during cultivation and domestication. However, the principles illustrated in terms of the domestication of cereals apply to other species. The differences are largely in details—for example, in leguminous and cruciferous crops indehiscent pods and siliques evolved instead of nonfragile rachises and fragile glumes. Other parallel evolutionary trends have been noted under circumstances of cultivation. Thus, harvesting and planting emerge as two key agricultural operations, because they often impose complex syndromes of interlocking selection pressures, some resulting simply from growing cultigens or domesticates under circumstances of cultivation, but others imposed by deliberate selection by humans. Combined natural and human selection seems likely to be a factor in modifying many characters, particularly those that have obvious effects on degrees of dormancy, photoperiodic response, time of harvest, organ size, and plant or seed size. A seed or vegetative propagule that does not germinate or break dormancy or start vegetative growth at precisely the right time in cultivation is likely to be rogued (eliminated) by humans during the growing season. If it escapes human selection until harvest and does not mature at precisely the time of harvest chosen by humans, it is still likely to be eliminated by natural selection. Maturation over a period of time is favorable for wild plants but is detrimental in cultivation because farmers prefer the obvious convenience of a single harvest. This is consistent with the stronger trend toward much more synchronous flowering and much more uniform whole-plant maturation of seed-bearing organs in modern cultivars than was the case with earlier cultivars or their wild relatives. It is also consistent with the trend toward apical dominance, larger inflorescences, and larger seeds, all of which contribute to improved convenience in harvesting. These trends suggest that human selection and natural selection may both have played consequential roles in the evolution of many traits, but do not settle the question of the relative roles played by each.

The environment of the cultivated field is very different from the environments of nature. The cultivator prepares a seedbed as favorable as possible for germination and establishment of the species under cultivation, and competition with other species is reduced as much as possible. However, stands are likely to be dense in cultivation, and competition among plants of the same species can be extremely intense in monoculture. Large seeds often produce larger and more competitive seedlings than smaller seeds; consequently, they are likely to contribute more offspring to the next generation than less vigorous seedlings. But plants that produce larger numbers of smaller seeds have a numerical advantage that may counterbalance the selection pressures favoring larger seeds. Thus, a balance is likely to be reached between size and the number of seeds. Similar, but apparently even more complex, interactions are likely to be a factor in achieving appropriately balanced expressions, especially for physiological traits that have little visible effect on the phenotype but may nevertheless have significant effects on adaptedness and survival ability. Such traits are often postulated to be controlled by interacting networks of loci-affecting enzymatic reactions that vary from habitat to habitat but are difficult to identify or to quantify individually or jointly. In some cases it appears that natural selection may be more discerning than farmers (or plant breeders) in identifying and preserving adaptively superior types, particularly if natural selection is allowed to operate for several to many generations in relevant environments.

## LANDRACES

Earlier we saw that the wild ancestors of cultivated plants are differentiated into genetically distinct populations as a result of natural selection and human-directed selection operating on genetic variability in the various ecogeographical locations in which the wild species had gained a foothold. We also saw that wild plants taken into cultivation become genetically altered in the process. The oldest evidence of undoubtedly cultivated materials comes from archaeological sites in the Near East, Central and South America, and South China, all dating to about 11,000 B.P. Although there is little direct evidence concerning the ways in which the earliest farmers propagated their plants during domestication, much can be inferred from archaeological records, from written records extending over many centuries (it is only in the last century or so that landraces have been largely replaced in developed countries by uniform true-breeding cultivars or hybrid varieties of closely controlled parentage), and from direct observation of the activities of present-day subsistence farmers who still practice traditional agriculture. The consensus is that even the earliest farmers were competent biologists who carefully selected as parents those individuals with characteristics that endowed the selected individuals with the ability to live and reproduce in the local environment, as well as with superior usefulness to local consumers. Moreover, when the progeny of the selected plants were planted to produce succeeding generations, natural selection (perhaps unperceived by humans) probably continued to adjust the ability of the population to prosper in local environments. Some modern subsistence farmers have been observed to select carefully for diversity of type, and others for uniformity of type. Whatever the reasons for their choices, certain components were saved because they appeared to be adapted to the local environment and because they appeared to be coadapted to each other. All landraces are or appear to be heterogeneous mixtures of genotypes that have mutually beneficial associations with each other and with environmental conditions. Dependability, especially the capacity to produce at least modest yields even when growing conditions are unfavorable, characterize modern landraces, and this has perhaps been the most important attribute of landraces through the centuries.

Many different species were domesticated at many different places. It is unlikely, however, that domestications were ever single events. Domestications were almost certainly ongoing processes carried out over many generations at many different places within each cradle of domestication. Thus, within the Fertile Crescent it is clear from the archaeological record that the major cereals were domesticated at many different localities, extending from the southern Levant north to Turkey and south and eastward to Iran, and that exchanges of materials from place to place within the Fertile Crescent took place at least sporadically. The genetic changes required for domestication were, for the most part, straightforward, and many changes (e.g., seed retention) depended, at least in part, on unplanned and/or perhaps unconsciously directed selection. Many improvements resulted from allelic changes at only one or two major loci, accompanied by the development at other loci of relatively simple modifier systems, as discussed earlier. Early in the domestication process many improvements were almost certainly both rapid and straightforward. The time required for domestication is not clear in any instance, but there is evidence that substantial local differentiation had developed within some cradles of domestication (such as the Fertile Crescent) by the time humans of the area had become dependent on agriculture for their own survival.

Soon after food production became firmly established in the Fertile Crescent, the Near Eastern suite of domesticates radiated to other parts of western Asia, North Africa, and southern Europe, thus starting the second phase of evolution under domestication. The Fertile Crescent package of domesticates reached Cypress and Greece by about 9000 B.P., southern Italy by about 8000 B.P., and Egypt and Pakistan also by about 8000 B.P. Not all of the items of the package became established in each new area; for example, Egypt was too hot for einkorn wheat. Most of the food production in these regions depended at first on the complete, or a nearly complete, group of Fertile Crescent domesticates. Evidently, the Fertile Crescent domesticates became established very rapidly in Greece, southern Italy, and Pakistan, primarily because these areas lie almost directly west or east of one another at nearly the same latitude, so that these localities have almost exactly the same seasonal variations in day length; photoperiod is the environmental factor that plays perhaps the largest role in fine-tuning flowering time. Temperatures, rainfall patterns, and other environmental features in these localities are also similar to those of the Fertile Crescent, but to a lesser extent than similarity in photoperiod.

We now contrast this mostly easy west-east diffusion of crops with the problems encountered in the third phase of evolution under domestication, when diffusion was into areas in which environmental circumstances often differed widely from those of the cradle of domestication. One well-documented case is that of the diffusion of the suite of Near Eastern crops northward through Europe and southward into Africa. The north-south diffusion was much slower than the east-west diffusion. It also featured many differences in the success of components during the diffusion process; for example, einkorn wheat became a much less important component at lower latitudes, whereas emmer and hexaploid wheats became progressively more prominent components in the prehistoric agriculture of northern Europe than in the agriculture of the Near East. Rye and oats also became more prominent during the northward diffusion into cooler areas, whereas lentils declined. Thus, transport to quite different environments exposed differences in the innate adaptability of different crops. Crops that were capable of adapting to new environmental circumstances quickly became established, and many less adaptable crops did not become established or took longer to become properly adapted for establishment.

Additional phases of the diffusion process started with the particularly active period of exploration, discovery, and colonization that followed the explorations of Columbus. In those extended phases of diffusion, very large numbers of crops were transported to new environments throughout the world. The introductions were landraces that had been shaped over the centuries by combined natural selection and human activities in countless farmers' fields in Asia, Europe, Africa, and the Americas. Vast numbers of types, adapted to a multitude of environmental circumstances and human preferences, had arisen; and when they were transported by humans to new environments, many mismatches occurred, resulting in numer-ous failures. However, some close matches were also found, leading to quick successes, and other introductions gradually became adapted to the environments to which they had been transferred. The materials introduced were usually hetero-geneous landraces and were thus readily susceptible to being shaped and molded, generation after generation, by the local environment, by local agricultural prac-tices, and by local human preferences, into new balanced and integrated mixtures of genotypes that gradually became better adapted to novel, specific local circum-stances. Thus, natural and human-guided selection were inextricably interwoven in the evolution of a great wave of newly developed landraces. Even though written descriptions of some of these landraces exist, it remains unclear precisely how much natural selection contributed to the improvement of adaptedness in the new environments. This is an important question, because recent studies of both wild and cultivated plant populations indicate that natural selection in itself is often successful in altering adaptively superior traits, traits that are often difficult for breeders to identify and quantify. The reason is that the effects of such traits on adaptedness nearly always depend on interactions among alleles of loci affecting two, and usually many more, biochemical reaction systems, each of which is often genetically complex within itself. The next section of this chapter is intended to provide a view of the intricacies of the adaptive changes that occurred, in hope that readers will more readily follow the details of the genetic basis for change when arguments regarding genetic change are presented in later chapters.

## GENETIC CHANGES ASSOCIATED WITH
## THE EVOLUTION OF ADAPTEDNESS

In this section we focus on studies of experimental populations of cultivated plants, among which barley Composite Cross II (CCII) has perhaps been examined most extensively and intensively. This population was synthesized in 1928 by pooling equal numbers of $F_2$ seeds obtained by selfing $F_1$ hybrid plants of the 378 possible pairwise intercrosses among 28 superior barley cultivars chosen to represent all of the major barley growing areas of the world (Harlan and Martini 1929). CCII has subsequently been grown annually on the experimental farm of the University of California, Davis, for more than 60 generations, following standard agricultural practices of the area. Subpopulations, derived from early generations of the Davis CCII population, have also been grown for shorter periods in a number of other locations to test the effects of different environments and agricultural practices (e.g., irrigated vs. dry-farm agriculture) on evolutionary changes. Each generation was allowed to reproduce by its natural mating system (~ 99% selfing and ~ 1% outcrossing), harvested in bulk at maturity without conscious selection, and the next generation was sown from a random sample of seeds from the previous harvest. In early generations ($F_2$ to $F_{15}$) randomly chosen plants of each generation were measured for several continuously varying traits (e.g., flowering time, plant height, number of seeds per spike, spike weight) and were scored for allelic state for several Mendelian loci that code for discretely recognizable variants (e.g., two-row vs. six-row spikes, rough vs. smooth awns, resistance vs. susceptibility to specific pathotypes of diseases). In later generations allelic frequencies for allozyme variants, DNA restriction fragment variants, microsatellite variants, and cytoplasmic DNA variants were also determined in plants grown from stored seed saved from earlier generations as new kinds of markers became available (ultimately, changes at more than 30 such marker loci were analyzed). In later generations populations were monitored less frequently, usually at five-generation intervals, in plants grown from stored seeds saved from previous generations. Analogous studies of other barley CC crosses and other cultivated species, including full outbreeders (corn), as well as mixed selfers and outcrossers (particularly sorghum and *Phaseolus* beans), were analyzed in similar ways. Because all generations were advanced with careful attention directed to avoiding human-directed selection of any kind, genetic changes in these populations can be presumed to be largely, and perhaps even entirely, due to selective pressures imposed by the agricultural environments prevailing at the locations where they were grown.

Considering the broad-based hybrid origins of the populations studied, it is not surprising that early generations were conspicuously more variable genetically than modern landraces or modern cultivars. Genetic variability was particularly apparent when families derived from seeds of randomly chosen single plants from the experimental populations were grown side by side in family rows. Under such conditions it was readily apparent that different families that were derived from different single plants varied continuously for many traits such as flowering time, inflorescence size and weight, and number of seeds per plant. Uniformity within families increased very rapidly in early generations for heavy inbreeders such as barley (~ 99% selfing), more slowly for weak outcrossers such as *Phaseolus* beans (3–5% outcrossing) and sorghum (3–10% outcrossing) and very much more slowly for corn (90% outcrossing, 10% selfing). In contrast, family-to-family differences in variability increased rapidly for barley, more slowly for *Phaseolus* beans and sorghum, and very slowly for corn. However, at the same time, the frequencies of families with extreme expressions of quantitative traits decreased. Moreover, all families, as well as all populations, became more uniform. Comparisons made over generations showed that very large directional changes occurred for some quantitative traits (e.g., number of spikes per plant; numbers of seeds per spike; spike weight; shorter, more compact spikes; and seed yield—all of which are direct components of fecundity). This was especially apparent in barley populations grown under consistently favorable conditions (irrigation, fertile soils) or consistently unfavorable conditions. Changes in these traits were often concurrent with increases in yield (determined by measuring the productivity of different generations in replicated small plot trials and by comparing the grain yields of early, intermediate, and late generations with each other and/or with several genetically highly uniform modern cultivars). Grain yields also increased steadily when the barley populations were grown in xeric habitats in which moisture stress was moderate to severe in most years. However, flowering time, height, and culm weight of the barley decreased steadily under such environmental conditions. The correlations between these traits and seed yield became increasingly negative over generations, indicating that more and more of the total biomass produced was being partitioned to seed production and less and less to vegetative growth under recurring conditions of moisture stress. Interestingly, most of the changes in the quantitative traits studied were erratic in direction under circumstances in which moisture levels were ample to adequate in some years but limiting in other years; selection pressures were clearly in opposite directions for many traits in years of ample rainfall versus years of moisture stress. Clearly, moisture was the crucial environmental factor in barley culture in semiarid regions of California. This was not the case with irrigated barley or with crops usually grown under irrigation (corn, *Phaseolus* beans, sorghum).

CCII and other analogous barley populations have also been studied intensively in respect to single-locus and multilocus allelic frequency changes over generations. These studies showed that alleles that were present in very high frequencies in the 28 parents of CCII consistently remained in high frequency into advanced generations of the experimental populations; these apparently are elite general-purpose alleles that do well in nearly all environments. Rare and very infrequent alleles also behaved consistently; they were nearly always quickly eliminated within a few generations. Rare alleles apparently produce defective phenotypes that do poorly in most environments. However, moderately frequent alleles of loci that were polymorphic in the parents of CCII nearly always behaved erratically. Sometimes specific alleles increased in frequency for one or a few generations and then reversed direction, usually in concert with the annual environmental fluctuations (particularly with shifts from conditions of drought to conditions of ample rainfall, and vice versa) that are commonplace in the ultra-Mediterranean climate of the Central Valley of California. These results suggest either that alleles of the marker loci have pleiotropic effects on one or more physiological processes that affect survivability and/or reproductive success, and/or that alleles of different loci that have physiologically significant effects are tightly linked to the marker loci. These hypotheses were examined through extensive progeny testing of highly isogenic lines carrying alternative alleles of the marker loci. Comparisons of alternative homozygous allelic variants revealed that isogenic pairs of many kinds of marker loci (e.g., loci coding for morphological, allozyme, RFLP, disease resistance vs. susceptibility variants) often had statistically significant effects on more than one quantitative trait, including traits affecting survival ability and fecundity. Thus, alleles of the marker loci tested often appeared to affect one or more quantitative trait in addition to the discretely distinctive (qualitative) characteristics from which the marker locus took its name. Perhaps more likely, but difficult to determine experimentally, is that alleles of entirely different loci occupy the same short segment of chromosome occupied by the marker locus.

It has been established experimentally that barleys from different eco-geographical regions of the world nearly always differ widely from one another in multilocus genetic structure, whereas the multilocus genetic structures of barleys from closely similar ecogeographical regions tend to be similar in their multilocus genetic structure. Studies of the population dynamics of the multilocus structure of wild as well as cultivated populations have revealed various features of multilocus genetic structure, among which the following appear to be particularly germane to plant-breeding procedures:

1. Statistically significant, nonrandom 2-locus, 3-locus, and 4-locus, or even higher-order associations of alleles of different loci, often developed within a few generations in each of the heavily inbreeding populations studied.
2. 2-locus and higher-order associations often coalesced within a few generations into statistically significant associations involving several alleles of several different loci.
3. Different populations grown in the same environments frequently developed similar multilocus structures.
4. Subpopulations of a single experimental population soon developed distinctly different multilocus genetic structures when subpopulations were grown in different environments.

Similarly, shifts in degree of quantitative trait expression (e.g., toward earlier maturity, shorter stature, and higher yield) occurred in xeric Californian habitats, and toward later maturity, larger kernel size as well as higher yields in fertile mesic (or fertile irrigated) habitats. These results led to two main conclusions regarding the effects of natural selection in heavily selfing populations: (1) that natural environmental perturbations play a major role in the evolutionary dynamics of such populations; hence, genetic studies based on only one or a few generations or only a few years of testing are unlikely to provide adequate characterizations of the causes or consequences of genetic changes, and (2) that natural selection alone can lead within a few generations to significant improvements in heavily selfing populations, improvements that apparently depend on development of favorable multilocus combinations of alleles in homozygous lines and the fixation of such homozygotes before they are broken up by segregation. When outcrossing is only slightly more frequent (3–10%), as in *Phaseolus* beans and sorghum, formation of favorable multilocus combinations and their subsequent destruction owing to hybridization among different genotypes, followed by segregation, apparently often canceled each other, and improved adaptedness and performance developed more slowly in any given environment. In predominant outbreeders such as corn (90% outcrossed, 10% selfed) all generations remained highly heterozygous and discernible multilocus population structure developed very slowly; seed yields also improved slowly. Modern corn breeders have circumvented this problem by developing homozygous inbred lines that, when crossed with other specific homozygous lines, produce particular multilocus combinations of alleles that often lead to high $F_1$ performance. Not surprisingly, as soon as any $F_1$ hybrid reproduces, segregation immediately destroys the favorable $F_1$ genotype, and the $F_2$ and later generations of the population become complex mixes of segregation products that are dramatically lower producing. Mating systems clearly play a major role in the effectiveness of the breeding methods that have been successful in plant improvement. Often small differences in amounts of selfing versus outcrossing greatly influence success; it is clearly advantageous for breeders to have precise information concerning the mating system of their materials. Consequently, we turn to methods of obtaining precise quantitative estimates of mating system parameters in the next chapter (Chapter 4) before proceeding to general patterns of breeding appropriate to crops with different reproductive systems (Chapter 5 and, especially, Chapters 12 through 17).

# FOUR

# *Mating Systems of Plants*

Wild plant species, as well as crop species, can be divided into three groups in respect to sexual mating and asexual reproductive systems: species that are predominantly self-pollinating, those that are largely cross-pollinating (outcrossers), and out-crossers that can be propagated vegetatively. The distinction between the first two groups is important, because breeding methods that are appropriate to self-pollinators are largely different from those appropriate to outcrossers. The main difference between the two groups stems from the differing effects of inbreeding versus outbreeding on the genetic structure of populations. Plants within individual populations of outbreeding species often carry different alleles at many loci, and their outcrossed as well as their selfed progeny segregate widely to produce highly heterogeneous progeny. Selfed progeny of largely outcrossing species always show decreased vigor and fitness; however, the crossing of different highly inbred and homozygous lines derived from outbreeding populations usually restores diversity and the crossing of carefully selected inbred lines sometimes results in remarkably superior performance. Hence, allelic diversity must be restored at the end of inbreeding programs. In contrast, most individuals within predominantly selfing populations are vigorous homozygotes whose progeny do not lose vigor on inbreeding, but sometimes show superior performance when intercrossed. Individual plants within long-inbred populations often differ in agricultural value and, as a consequence of their homozygosity, superior individuals selected from such populations usually breed true and often serve as the parents of vigorous true-breeding varieties or superior $F_1$ hybrid varieties. The general pattern of the breeding program appropriate to any plant species is therefore determined, in large part, by its mating system.

In addition to their influence in determining the general features of breeding programs, mating systems also play a central role in determining the specific breeding procedures feasible with a given species. Thus, the ease with which selected individuals can be selfed, and/or controlled hybrids can be made between selected genotypes, is often critical in determining not only the general features but also the details of execution of the most appropriate breeding program for any particular species. A detailed discussion of the systems of mating used by plant breeders and the genetic consequences of these mating systems is included in Part III of this book.

The type of program that has been most successful with corn—namely, exploiting heterosis by growing exceptionally heterotic $F_1$ single-cross hybrids between carefully selected inbred lines (single-cross hybrids) in commercial production—is feasible because selfing is easy, because many modern selfed lines are easy to maintain and reasonably productive, and large-scale hybridizations between pairs of selfed lines can be achieved by interplanting pairs of inbred lines and, at flowering, detasseling one or the other of the inbred lines mechanically, or by taking advantage of one or another sterility mechanisms to obtain hybrids in mixed plantings. Alfalfa is also an outbreeding species, but one in which both selfing and controlled outcrossing are difficult. It has, nevertheless, been possible to exploit heterosis in alfalfa but usually in different ways than those used with corn, for example, by developing synthetic varieties. Synthetic varieties are those synthesized by open pollination between a number of genotypes that are known from previous testing to be heterotic in all hybrid combinations. In recent decades, use of male sterility has made it possible to apply single-cross methods to sorghum, a self-pollinated species, and to a number of other predominantly self-pollinated species. Further, when vegetative parts of a plant can be used to produce clones, many additional patterns of breeding become possible.

Once superior varieties have been developed, it is necessary to protect their genetic integrity during seed increase and distribution to farmers. A few species are so highly self-pollinating that measures to protect against outcrossing are unnecessary even when many different genotypes are grown in close proximity in breeding nurseries and/or seed increase nurseries. However, outcrosses occur often enough in most predominantly self-pollinated agricultural species that some form of pollination control is required.

Clearly, mode of reproduction affects so many aspects of plant breeding that a discussion of the main types of plant reproductive systems is called for prior to discussing breeding methods themselves. In this chapter we consider some of the more common systems by which cultivated plants reproduce themselves. The impact of these reproductive systems on plant-breeding procedures are considered in greater detail throughout the rest of this book.

## DEVICES FOR POLLINATION CONTROL

Devices for the control of a breeding system take a variety of forms in both wild and cultivated plants. The most common natural inbreeding devices feature cleistogamous (closed) flowers. In some grasses, not only are the florets themselves closed but they are further enclosed between the culm and the leaf sheath and the inflorescences often do not emerge from the leaf sheath. Seeds from such doubly protected flowers thus nearly always result from self-pollinations. In other grasses, the inflorescences emerge from the boot formed by the leaf sheaths, but the florets open only after the anthers have burst. Similarly, the anthers of some species (e.g., lettuce) usually rupture and the stigma is normally pollinated before the flower opens. Such species are naturally almost entirely self-pollinated; however, even the limited number of outcrosses that occur produce segregants that have been important in the evolution of such species in nature and in cultivation. In barley, wheat, and many wild grasses the anthers burst more or less coincidentally with the emergence of spikes from the boot; the borderline cleistogamy of these species thus permits low levels of outcrossing that can lead to continuing evolutionary change. The flowers of subterranean clover are generally borne underground, which is equivalent in its effect to morphological cleistogamy; nevertheless, rare outcrossed seeds have been reported in this forage legume. Pollination in the cultivated tomato follows the opening of the flower, but the stamens form a cone, enclosing the stigma in such a way that self-pollination is almost ensured. There is, however, some variation in the length of the style, which affects the position of the stigma within the cone of anthers, with the result that the amount of outcrossing differs from genotype to genotype. It should be recognized that the effectiveness of different natural inbreeding devices is subject to modification by both genetic and environmental conditions, and that levels of outcrossing in a given species sometimes vary substantially from one genetic background to another and/or with environmental conditions in different seasons and/or locations. Consequently, it is desirable for breeders to determine the amount of outcrossing that occurs in their particular breeding materials in their local environments. Note that none of the aforementioned inbreeding devices requires genetic diversity among individuals within populations.

There are also a number of outbreeding devices that do not depend on genetic diversity among individuals. The most common of such devices feature the opening of flowers (chasmogany) before anthers and stigmas mature, and the maturing of the anthers and stigmas of single flowers at different times. Corn, carrots, and raspberries are often protandrous (pollen shed first), whereas avocados and walnuts are protogynous species (stigmas receptive first). In corn, male and female gametes are separated in space. Pollen is produced in the tassel located at the top of each plant, and ovules are produced in the ear located on the culm well below the tassel; this promotes outcrossing because wind disperses pollen in various directions from the tassel. Delay in silking is often deleterious in corn, especially under modern high-density planting and high-yield conditions, but breeders have been successful in developing varieties with synchronized or protogynous flowering, thus reducing the deleterious effect of delayed silking (Troyer and Brown 1976). Alfalfa has a different outbreeding device. In this species the stigma does not become receptive until a protective membrane enclosing it is ruptured. This is accomplished by "tripping." In alfalfa the stamens and stigma grow inside the keel, which holds them under considerable tension. When this tension is released by mechanical pressure, often supplied by visiting bees, the stigma snaps against the standard, the membrane is ruptured, and pollen is released. Some of the pollen adheres to the hairy body of the bee, serving to pollinate successions of flowers next visited.

Although there have rarely been measurements made of the precise effects of such outbreeding devices on suppressing inbreeding, it is well established in plant breeding lore that these devices can be very effective. Beets, which feature protandry, are highly outbred; the same is true of red clover. The high degree of outcrossing in beets and red clover may not, however, be due to dichogamy alone. Red clover, for example, is highly self-incompatible in addition to being protandrous, and the intricate entomophalous pollination system of alfalfa is reinforced by some sort of incompatibility system. Further, embryos derived from selfing are less likely to survive than embryos produced by outcrossing. This phenomenon, which enhances the apparent level of outcrossing, has also been reported in many conifers as well as in many angiosperms. If little is known about the effectiveness of the previously mentioned outbreeding devices, even less is known about their genetic control. Perhaps the most studied species is corn, which is highly outcrossed. Nearly all embryos (usually ~ 90%) result from biparental matings primarily resulting from separation of male and female flowers in space, sometimes combined with protandry or protogyny in modern corn varieties.

The most obvious of the outbreeding devices that require genetic diversity among individuals is, of course, dioecy (staminate and pistillate flowers borne on different individuals). Dioecy is clearly an outbreeding device because, when absolute, it obviously prohibits selfing, the most intensive form of inbreeding; however, even perfect dioecy does not prohibit full-sib or other less severe forms of inbreeding. Dioecy, the great mating system of animals, is uncommon in higher plants, perhaps because it is wasteful of gametes in nonmobile organisms. Among cultivated plants some of the more important dioecious species are date palms, castor beans, hemp, hops, spinach, papayas, and asparagus. Some individuals of these species produce hermaphroditic as well as staminate and pistillate flowers, and when this is the case, at least some selfing is possible. The genetic control of dioecy in plants has received considerable attention over the years, and the genetic control of dioecy is reasonably well understood in many species (review in Westergaard 1958).

Self-incompatibility is far more important than dioecy as an outbreeding device in cultivated plants. There are two general schemes of self-incompatibility in higher plants: (1) gametophytic or haplo-diplo schemes in which incompatibility depends on the genotype of the gametophyte and (2) sporophytic or diplo-diplo schemes in which incompatibility is impressed on the gametophyte by its sporophytic parent. In gametophytic systems no seeds are produced on selfing, but on outcrossing either all of the pollen grows (full compatibility), one-half grows and one-half fails (half compatibility), or none of the pollen grows (full incompatibility). With sporophytic systems no seeds are produced on selfing, but with crossing either none or all of the pollen grows.

## ASEXUAL REPRODUCTION

Reproduction by asexual means is common in higher plants. The most familiar methods by which farmers have reproduced cultivated plants asexually are via corms, bulbs, rhizomes, stolons, tubers, or other vegetative organs or tissues. Plants that have been routinely propagated in agricultural practice for centuries by these means, and in more recent times by budding or grafting, include nearly all fruit and nut trees, strawberries, blackberries, raspberries, grapes, pineapples, some field crops, including potatoes, sugar cane, yams, and sweet potatoes, and many ornamental species. Vegetative reproduction leads to perpetuation, with great precision, of the genotype of the individual being reproduced. In many cases vegetative reproduction makes it possible to produce indefinitely large numbers of genetically identical individuals over large numbers of generations. Thus, the breeder can take

advantage of outstanding individuals that appear at any stage in a breeding program, immediately fixing their superiority through vegetative reproduction. In
addition to allowing perpetuation of desirable recombinants following sexual
reproduction, vegetative reproduction also permits perpetuation of occasional
desirable mutants (sports) that may occur in vegetatively reproduced varieties;
vegetative reproduction also allows immediate elimination of undesirable mutants.
Recent advances in methods of growing cells, tissues, organs, or whole plants in
vitro (tissue culture, in the broad sense) have greatly expanded opportunities for
taking advantage of vegetative reproduction in plant breeding.

In addition to the types of vegetative reproduction commonly used in the
horticultural arts, there are a number of types of apomictic reproduction that
feature the external trappings of sexual reproduction but with the omission of
fertilization, and often with the omission of meiosis as well. The main effect of
apomixis is to increase the proportion of individuals that resemble the maternal
parent in the next generation. The progeny of apomictic individuals are frequently
variable genetically; apomixis consequently finds relatively little use in plant-
breeding practice.

## DETERMINING MODE OF REPRODUCTION AND PREVALENCE OF NATURAL CROSSING

Most cultivated species have been under close observation for hundreds of years,
and the general features of their reproductive systems are reasonably well known.
However, information about mating systems is incomplete to varying degrees for
all species, particularly regarding the frequency with which natural crossing occurs
over the range of environmental conditions in which any particular species is
cultivated. Observations and experiments to determine whether a species is predominantly self- or cross-pollinated in any particular set of environmental circumstances are usually fairly simple and straightforward. Examination of floral
structure is an obvious first step, and for some species the presence of outbreeding
devices (e.g., staminate, pistillate, protandrous, protogynous flowers) provides
presumptive evidence that a species is predominantly outcrossing, just as cleistogamy provides presumptive evidence of predominant selfing. Usually, the next
step is to isolate single plants and determine the number of seeds they produce
when forced to self-pollinate. Isolation in space is the preferred procedure, because
isolation by bags, cages, or other restraining devices introduces the possibility of
imposing environmental conditions adverse to seed production. Failure of a single
plant to set any seed at all, or poor seed set in isolation, provides evidence that the
species is cross-pollinated. The reverse is not necessarily the case, because many
normally cross-pollinated species (e.g., corn), are highly self-fruitful. The next
question that arises concerns the amount of natural crossing that occurs within
populations, or when different genotypes are grown at various distances from one
another. Earliest estimates of the amount of outcrossing were based on opportunities that arose occasionally when morphologically distinct varieties that produce
recognizable hybrids had been grown in proximity to each other under various
circumstances—for example, in adjacent commercial fields or in adjacent plots in

breeders' nurseries. The outcomes of many such "opportunities" were incorporated into plant-breeding lore, and it became widely accepted, and appropriately so, that outcrosses occur only rarely in some species (e.g., barley, wheat) and then nearly always as a result of pollen migrations between plants located close to each other. In other species (e.g., corn) it came to be accepted that although the majority of seeds result from outcrosses among plants of the same population, pollen migrations can occur over substantial distances. However, such traditional methods do not necessarily provide reliable quantitative measures of the success of matings between genetically different individuals in populations, and the information such methods provide is often inadequate for analyzing genetic transmission at the population level. More recently deliberate experiments based on increasingly efficient experimental and statistical procedures have been used to obtain precise quantitative measures of relevant mating system parameters.

The pioneering study was by Jones (1916), who planted equal numbers of homozygous dwarf ($dd$) and homozygous standard ($DD$) tomato plants in alternate rows, harvested seeds from the single-locus dwarf plants, and counted the number of standard plants ($Dd$ outcrosses) in the progeny of dwarf plants. Jones noted that the outcrosses observed in the progeny of dwarfs do not represent all of the crossing that might have occurred; it is reasonable to assume that there would be an equal chance that dwarfs might be fertilized by pollen from other dwarfs and that such crosses would not be detected. Thus Jones recognized that it was necessary to compensate for cryptic outcrosses, and he multiplied the observed proportion of outcrosses by 2 to obtain an estimate of the outcrossing rate. Jones's intuitive estimator of the outcrossing rate, $t$, can be shown to be the maximum likelihood estimator $t = h/p$, in which $h$ is the frequency of recognizable outcrosses in the progeny of recessives and $p$ is the frequency of dominant alleles in the pollen pool. For the next half century most experiments conducted to estimate mating system parameters were modeled on that of Jones. We turn now to experiments with two extensively studied species, barley (an inbreeder) and corn (an outbreeder), to illustrate the results obtained using modifications of the approach developed by Jones.

In one set of experiments designed to measure outcrossing rates among closely adjacent plants of barley, seeds of a pollen and a recipient parent were mixed in 1:1 proportions and seeded in an isolated block of rows, which were grown at densities such that all or nearly all of plants in the same row were in direct physical contact with one or more other plants in the same row. The number of outcrosses observed in approximately 1,000 progeny of recessive plants that had been scored in separate experiments conducted in several locations and years varied from 3 to 19, leading to estimates ranging from $t = 0.6\%$–$3.8\%$. In other experiments, distances between pollen and recipient parents were varied in a number of ways. In one experiment with barley, direct physical contact between pollen and recipient parents was largely eliminated by spacing plants alternately 30 cm apart within and between rows. Under this planting plan mean outcrossing rates fell to $\sim 0.2\%$ or approximately one-fifth the rate observed in dense stands. The mean outcrossing rate was also $\sim 0.2\%$ when donor and recipient parents were planted in pure stands in adjacent rows spaced 30 cm apart. When donor and recipient barley parents were planted two rows (60 cm) apart, separated by a row of wheat, or three rows (90 cm) apart, separated from each other by two rows of wheat, mean outcrossing rates fell to ~ 0.10% to ~ 0.05%. When potential parents were separated by 3 m up to 30 m, fewer than one outcross was observed per 10,000 recessive plants scored. Opportunities to estimate rates of pollen migration have also arisen occasionally when adjacent parts of large commercial fields (e.g., fields of barley) were seeded to different varieties. Usually fewer than one outcross was observed per 10,000 plants scored when the nearest pollen parent was 1 m to 10 m distant from the recipient parent. Outcrosses were also observed very rarely when the separation was 20 m or more, even though 10,000 to 20,000 progeny of recessive parents were scored. The greatest distance at which an outcross has been observed in barley in California was 60 m; in this particular case only one outcross was observed in > 20,000 individuals scored.

It became clear from experiments such as these that distance is the single most important factor affecting rates of successful pollen migrations in barley and many other predominantly self-pollinating plants. However, these experiments also established that overlap in flowering time is an important factor as well. Varieties of heavily selfing cereals that flower concurrently outcross at much higher rates than varieties that differ by as little as 2 or 3 days in time of maximum coincident spike emergence. There is also evidence that some varieties of barley with nearly identical flowering times differ significantly in their effectiveness as pollen and/or recipient parents, but reasons for the differences were not readily apparent. Prevailing wind direction also influences pollen migration rates. Not surprisingly, more outcrosses were observed downwind than upwind, including cases (e.g., in *Phaseolus* beans) in which insects (thrips) were the pollen vectors. Moreover, outcrossing rates have nearly always been found to be lower in more arid locations and in years when conditions were unusually dry and/or hot during flowering.

The results of experiments such as the aforementioned thus support the conviction, firmly ingrained in plant-breeding lore, that outcrossing rates are usually very low for species with cleistogamous flowers. This conviction has led many breeders, seedsmen, and gene-bank managers to the conclusion that there is little need to protect stocks of such species from contamination by pollen migration. However, these careful experiments show that potentially significant amounts of outcrossing sometimes occur even in plants with cleistogamous flowers, especially over short distances. Thus, taking steps to protect the genetic integrity of materials that are particularly affected by contamination resulting from pollen migration (e.g., pedigreed stocks of seedsmen, certain research materials) should be a matter of routine. Fortunately, such protection is easily accomplished with many species; ordinarily, all that is required is relatively short-distance isolation. Obviously, exceptional care should also be taken to prevent seed contamination by volunteer plants in nurseries or through mixing of seeds during harvesting or seed-cleaning operations, or resulting from cryptic intercrossing within populations.

The approach originated by Jones has provided much information that has proved useful to breeders of predominantly self-pollinating species. The method has, however, been less useful with outcrossing species for two main reasons: (1) stocks homozygous for recessive markers are often not available in outcrossers, and (2) when such stocks have been available, there have often been differences in the effectiveness of the stocks as male and female parents, which can lead to biased estimates of the normal mating system. In corn, random mating has usually been assumed; however, in several studies of mixed polycross plantings of corn stocks that were homozygous for recessive markers, it was found that differences in time of silking and tasseling of the marker stocks led to major departures from random mating (Guttierez and Sprague 1959). The usefulness of the approach originated by Jones was subsequently greatly increased by two developments. The first was the formulation of estimators of the amount of outcrossing ($t$) versus selfing ($s = 1 - t$), from populational data, thus extending the use of the Jones approach from contrived experimental populations to natural or other populations in which allelic frequencies are unknown (Fyfe and Bailey 1951). The second development featured the use of allozyme and other electrophoretically detectable genetic variation to determine the frequency of homozygotes and heterozygotes in the progeny of single maternal individuals and the formulation of estimators of population parameters and their variances from such frequency data (Brown and Allard 1970). Electrophoretic data are particularly useful for this purpose, because populations are often polymorphic for numerous different electrophoretically detectable loci, and such variants are often codominant—so all genotypes (homozygotes and/or heterozygotes) can be identified from their electrophoretic phenotype without progeny testing. The experimental procedure involves (1) collecting seeds from randomly chosen plants in the population under study, (2) germinating several seeds from each maternal plant under favorable conditions to minimize the possible effects of selection, and (3) assaying the seedlings electrophoretically for several genetic loci to determine the genetic composition of progeny arrays derived from single maternal individuals. The genotypic composition of each progeny array depends on the genotype of the maternal parent, allelic frequencies in the pollen pool, and the mating system ($s$ and $t$). For diallelic loci failure to observe an $a_2 a_2$ homozygote in a progeny array identifies $a_1 a_1$ maternal individuals with increasing certainty as progeny size increases. Similarly, failure to observe an $a_1 a_1$ individual in a progeny identifies $a_2 a_2$ maternal individuals. The probability of correct identification is a function of allelic frequencies and the outcrossing rate. When neither $p$ or $q$ (frequency of alleles $a_1$ vs. $a_2$) is not nearly 0 or 1, three or four progeny are adequate for correct identification (P ~ 0.99) in moderately heavily selfing populations ($t < 0.10$), whereas in outcrossing populations ($t > 0.90$) eight or nine progeny per array are adequate for correct identification. Data from progeny arrays thus allow direct estimates to be made of the genetic composition of the array of genotypes of the reproducing adults in the parental generation (generation 1), the genetic composition of each progeny array (generation 2), allelic frequencies ($p$ and $q$) in the pollen pool, and the mating system parameters ($s$ and $t$) for each locus. Maximum likelihood estimates of $s$ and $t$ are obtained by iteration of the array of observed progeny genotypes, following the familiar Newton-Raphson method. Single-progeny conditional probabilities for the diallelic case are given in Table 4-1; the diallelic model is readily expanded to include three or more alleles/locus. Enzyme loci with codominant alleles have been used to obtain quantitative estimates of the evolution and various other aspects of mating system parameters in many plant species (review in Epperson and Allard 1984). The first study of an outcrossing species based on allozymes was of two Californian populations of corn, each synthesized from intercrosses among six inbred lines, followed by two cycles

TABLE 4-1  Single-progeny conditional probabilities for one locus with two alleles

| MATERNAL PARENT | PROGENY | | |
|---|---|---|---|
| | $a_1a_1$ | $a_1a_2$ | $a_2a_2$ |
| $a_1a_1$ | $1 - qt$ | $qt$ | 0 |
| $a_1a_2$ | $(s + 2pt/4)$ | $\frac{1}{2}$ | $(s + 2pt/4)$ |
| $a_2a_2$ | 0 | $pt$ | $1 - pt$ |

$p$ and $q$ = frequencies of alleles $a_1$ and $a_2$ in the pollen pool, $p + q = 1$.
$s$ = proportion of ovules self-fertilized.
$t = 1 - s$ = proportion of ovules from random outcrosses.

of reciprocal recurrent selection and one or two generations of open pollination (Brown and Allard 1970). On the basis of parentage, one of the populations was expected to be polymorphic for two alleles each of seven loci, and the other for two alleles for each of nine loci. Ten seedlings from 50 open-pollinated ears in one population and 60 open-pollinated ears in the other population were assayed electrophoretically. Estimates of $t$ varied from 0.93 to 1.10 in one of the populations and from 0.90 to 1.12 in the other population. None of the 16 estimates of $t$ was significantly different from unity, and mean values of $t$ for the two populations (1.00 and 0.97) were also not significantly different from 1.0. However, 10 among the 16 estimates were smaller than 1.0, which suggests that there may have been fewer heterozygotes than expected under random mating. Thus, although the data yielded estimates of outcrossing nearly equivalent to random mating, they clearly did not eliminate the possibility that the deficiency in heterozygotes, if real, may be due to the invalidity of one or more of the assumptions made in formulating the mixed-mating mode.

Subsequently, methods have been developed for estimating mating system parameters that are more efficient than single-locus estimation and that are also less sensitive to invalidity of the assumptions made in formulating the mixed-mating model (Shaw et al. 1981). The multilocus estimator takes advantage of the fact that each seedling has been classified for several loci: one of the two gametes of each member of any family obviously must have come from the maternal parent, whereas the second gamete may also have come from the maternal parent as a result of self-fertilization, but it may also have come from a different plant as a result of outcrossing. If the second gamete carries an allele that differs from the two alleles carried by the maternal parent at any one among the several loci for which each family has been classified, this single alien allele immediately marks that individual as having arisen from an outcross. Thus, multilocus classification correctly identifies as outcrosses some progeny individuals that would have been incorrectly classified as selfs in single-locus estimation. In a study of an open-pollinated population of corn, 370 seedlings from 40 open-pollinated ears were classified for eight allozyme loci. Single-locus estimates of $t$ for the eight loci varied from 0.778 to 0.996; four of the eight estimates were significantly smaller than unity. Thus, these single-locus estimates, in common with single-locus estimates that had been made in many different species, varied substantially over loci. It is possible that such differences between loci reflect the large variances that are common in single-locus estimation; however, it is also possible that invalidity of one or more of the assumptions made in formulating the mixed-mating model may be responsible. A multilocus estimate of $t$, made from the same data set, gave $t = 0.910$ with a standard error of 0.019. This value is slightly, but not significantly, higher than the mean value $t = 0.880$ of the eight single-locus estimates. Closely similar results were obtained in a study of the mating systems of four Nebraska populations of corn developed by mass selection (Kahler et al. 1984b). Single-locus estimates of $t$ fluctuated widely from locus to locus in each of these four populations; the mean of the single-locus outcrossing estimates was $t = 0.89 \pm 0.02$, whereas the mean multilocus estimate for the four closely similar populations was $t = 0.90 \pm 0.01$.

One of the major assumptions made in formulating the mixed-mating model is that allelic frequencies in the pollen pool are uniform over the area sampled. Two possible causes of heterogeneous distributions are (1) microhabitat selection, causing similar genotypes to be clustered within populations; and (2) the tendency of seeds to fall and grow near their maternal parent, causing relatives to be clustered. Such clustering can cause the frequency of homozygotes to be overestimated if sampling is carried out with the distribution assumed to be homogeneous. Clustering can also lead to inbreeding, because neighbors are more likely to be relatives than random members of the population and, owing to proximity, they are more likely to mate, causing single-locus estimates to be further biased downward. In all of these experimental maize populations, the seeds of the parental generation were harvested in mass, thoroughly mixed during harvesting and cleaning, and a random sample of seeds was planted each year. Hence, it is unlikely that the distribution of genotypes within each population and the distributions of alleles in each pollen pool differed in the various areas sampled. It consequently seems unlikely that the variability in single-locus estimates of $t$ in these studies resulted from heterogeneity within the experimental areas; statistical inefficiency appears to be a more likely candidate. Brown and Allard (1970) and Brown (1975) have discussed statistical efficiency, directing attention particularly to allocation of resources within and between family arrays. Extensive experience indicates that estimates of $t$ decrease in variability as the numbers of loci assayed are increased, presumably because differences among loci cancel out. This suggests that as many loci should be assayed in estimating mating system parameters as is consistent with allocation of resources among competing projects.

There are three additional assumptions made in formulating the mixed-mating model: (1) the probability of an outcross is independent of the maternal genotype, (2) selection does not intervene between mating and the time of determination of progeny genotypes, and (3) no associations exist between alleles of different loci (i.e., absence of disequilibrium). Invalidity of any of these assumptions can influence both single-locus and multilocus estimates, but, in general, multilocus estimates are expected to be less affected. The effects of these three phenomena will usually be confounded, and they often cannot be disentangled from population data. Measurement of their individual effects therefore requires experiments designed specifically to isolate appropriate degrees of freedom. Experiments to test the aforementioned assumptions apparently have not yet been carried out.

However, the similarity and homogeneity of means of single-locus and multilocus estimates of *t* in the experiments with corn and several other outcrossing species suggest that the assumptions of the mixed-mating model have usually not been violated substantially.

The results of the experiments with corn can be summarized as follows. First, about 90% of offspring in corn result from outcrosses and about 10% from selfing; the selfing leads to a significant excess of homozygotes and to deficiencies of heterozygotes, on average, in the progeny arrays of maternal plants immediately after the mating cycle. Second, genotypic frequencies in adult-generation maternal plants, inferred from electrophoretic assays, were consistently in accord with expectations under random mating. This suggests that selection favoring heterozygotes over homozygotes may have occurred at one or more intervals between the early seedling stage and the adult stage of the life cycle, thus correcting the imbalance between heterozygotes and homozygotes that resulted from the selfing. Therefore, even though selfing in corn leads to significant increases in the numbers of homozygotes immediately after the mating cycle, the effects of selfing do not accumulate over generations. The mating system of corn thus generates great genotypic variability, whereas selection operating at various intervals between mating cycles preserves adaptively superior genotypes and eliminates adaptively inferior genotypes. Studies of other predominantly outcrossing species (e.g., rye grass, Douglas fir, and lodgepole pine) have led to similar conclusions for these species.

The mixed-mating model has also proved useful in studies of both wild and cultivated populations of various predominantly selfing species such as barley, oats, and small fescue. Small fescue (*Festuca microstachys*), a wild rangeland grass very widely distributed in an exceptionally diverse range of habitats in western North America, appears to be one of the most highly self-pollinated species of plants. The lemma and palea of this grass enclose the stamens and stigma so closely that early observers and investigators estimated that fewer than 1 fertilization in 1,000 was likely to result from outcrossing. Precise estimates have confirmed that outcrossing rates in this species are very low. Nevertheless, populations of small fescue adapted to distinctly different habitats all appeared to be nearly as variable for quantitatively inherited characters as populations of predominantly outcrossing species. This apparent dilemma led to a series of mixed-mating–model studies of this species to obtain precise estimates of mating system parameters and the extent of genetic variability for discretely recognizable allozyme variants, as well as quantitatively inherited tracts. It was found that nearly all populations of small fescue are made up of several to many different kinds of highly homozygous individuals, the great majority of which differ from each other at many allozyme loci. Thus, many of the outcrosses that occur must produce a different kind of multiple heterozygote, each of which segregates on self-fertilization, ultimately producing many highly homozygous but genetically distinct genotypes, many of which persist in the population in which they arise. Studies of many other self-pollinating species, both wild and cultivated, have shown that extensive genotypic variability, present in the form of numerous distinctive homozygous lines, is a usual feature of within-population variability in most populations of such species and apparently the cause of much of the variability in such species, contrary to opinions frequently expounded in many evolutionary and plant-breeding treatises.

The results of detailed studies of many outcrossing, as well as many selfing, species thus support the premise that the mating system plays a central role in the generation of genetic variability. However, experiments discussed also suggest that other factors, especially selection affecting different homozygous multilocus genotypes differentially at various stages of the life cycle, also play major roles, often leading to interactions that affect the course of evolutionary change in complex ways. The dynamics of population change are discussed in detail following the appropriate background information concerning basic population and ecological and evolutionary genetics developed in Parts II and III.

# FIVE

# *Overview of Plant Breeding*

The purpose of this chapter is to provide a bird's-eye view of the genetic and evolutionary principles on which breeding plans are based. Plant breeding is concerned, in considerable part, with three main ideas: (1) the expressions of genes, (2) the behavior of genes in populations, and (3) the evolution of breeding populations by allelic substitutions under natural selection, supplemented by artificial selection imposed by breeders. More complete descriptions of the basic genetic and evolutionary principles on which successful breeding plans depend are given in Part II, "Biological Foundations of Plant Breeding," and the particulars of breeding operations themselves follow in Part III, "Modern Breeding Plans."

## GENOTYPE AND PHENOTYPE

It seems appropriate to start this broad conceptual overview of genetic principles on which plant-breeding plans are based with perhaps the most basic concept of genetics and plant breeding—namely, the relationship between genotype and phenotype. Within 3 years of the rediscovery of Gregor Mendel's trailblazing paper of 1865, Wilhelm Johannsen (1903) of Denmark showed that the genotype (the sum total of the genetic information of an organism) determines potential for development, whereas the environment in which the organism lives sets the developmental scheme actually carried out. Mendel and Johannsen thus showed that genotype and environment are jointly responsible for the phenotype ultimately realized. The breeder's task is to assemble within populations combinations of alleles of many genetic loci that lead to superior multilocus genotypes and, hence, to superior phenotypes in the environment envisioned for the potential new variety. Although visual inspections of phenotypes by breeders can often be useful in eliminating plainly undesirable phenotypes in variable breeding populations, the genotypes that are ultimately likely to be successful can less often be identified by visual inspection alone. Nearly always, multiple measurements of various aspects of the phenotype (e.g., yielding ability, various components of quality) must be made in several relevant environments to determine whether any genotype (or group of similar genotypes) that survive into the late stages of a breeding program will be suitable both agriculturally and in satisfying human needs. Furthermore, consumer preferences often change, so breeders must be prepared to adjust their initial goals, often on short notice. Such shifts do not necessarily change selection practices during the course of a breeding program itself, but may alter only the evaluation tests that are applied to the products of the breeding program. Breeders are therefore applied evolutionists, who choose apparently suitable parents, hybridize these parents, and attempt to develop populations containing seemingly desirable genotypes, using suitable combinations of natural and artificial selection in guiding allelic substitutions toward hopefully realizable goals.

Before exploring the rules for guiding allelic substitutions, it seems appropriate to consider the terminology of expression for the alternative alleles that may be encountered at any single genetic locus. First, any single locus can be known to exist with certainty only if at least two alleles with distinguishable phenotypic expressions have been shown by segregation tests to be responsible for the phenotypic expressions of the locus. An allele is said to be fully dominant when the phenotype of one homozygote (say $a_1 a_1$) and the heterozygote (say $a_1 a_2$) are indistinguishable from each other phenotypically. Alleles are said to be codominant when both are distinguishable in the heterozygote, partially dominant when the heterozygote is intermediate between the parental types, and overdominant when the phenotype of the heterozygote lies outside the range of one or the other of the parental types.

A number of additional points are important here. Breeders are usually concerned with the leading effects of genes—for example, whether a specific allele leads to tall plants or to short plants or to superior versus inferior ability to survive in a given environment. Consequently, breeders often chose to make their measurements on some familiar and convenient linear scale, such as in centimeters in the case of plant height, or the percentage of individuals that are able to survive in a given environment in the case of survival. But such characteristics often appear to be the end products of intricate chains of developmental sequences that may be expressed on nonlinear scales. Accordingly, breeders sometimes convert convenient linear-scale measurements to some nonlinear scale, such as a log or a square-root scale, that may shift allelic expressions in ways the breeder believes will more realistically represent gene expressions at the phenotypic level. Moreover, the phenotype as a whole often depends on the "background" genotype of a population; hence, when the scale of measurement is changed for one characteristic, the change may shift the expression of this characteristic relative to the expression of other characteristics. Further complications are likely to develop because background allelic and genotypic frequencies and expressions may shift with environmental variations from year to year and/or from place to place. Thus, breeders often attempt to identify and take advantage of various measurement scales in their efforts to develop superior agricultural populations from intercrosses between pairs, or perhaps even among many promising parents, usually parents with compensating strengths and weaknesses. Yet what actually happens in most hybrid populations is likely to be so complex that it is rarely realistic to expect to know precisely what the genetic situation is at any stage in the evolutionary processes that occur from generation to generation or from place to place. Fortunately, exact knowledge is unnecessary because the usual experience, particularly with populations started from intercrosses between two or more parents with differing strengths and weaknesses, is gradually increasing progress over generations, even in the absence of assistance from the breeder. It has been usual to attribute such progress to overdominance such that the effects of different alleles of any given locus are more extreme in the heterozygote $(a_1a_2)$ than in either homozygote $(a_1a_1$ or $a_2a_2)$. However, there is increasing recognition that single-locus overdominance presents major operational problems and that overdominance rarely exists at the single-locus level. Moreover, pseudo-overdominance, owing to epistasis at the two-locus and multilocus levels, when combined with tight linkage, is less beset with operational problems than single-locus overdominance. Thus, epistasis is now increasingly held to be responsible for most of the so-called heterotic situations that have been observed in both inbreeding and outbreeding species. In the two-locus epistatic case, tightly linked loci *a* and *b* are now commonly replaced in breeders' thinking with a single functional locus, such that $c_1 = a_1b_2$, $c_2 = a_2b_1$, and $c_1c_1 < c_1c_2 > c_2c_2$. Increasingly, population genetic analyses have established that epistasis, functioning to assemble and maintain favorable genotypes involving two, or often more, loci, is an important mechanism widely involved in the evolution of various natural and cultivated plants species, including both inbreeders and outbreeders. However, the development of population-genetic ideas, even in the simple cases of the behavior of only a pair of alleles of a single Mendelian locus in populations, developed very slowly until the last decade or so of the twentieth century.

Mendel himself was the first to consider the problem of a single pair of alleles at the population level. After establishing the 1:2:1 ratio for homozygotes of one kind (*AA*), for heterozygotes (*Aa*), and for homozygotes of the opposite kind (*aa*), in the $F_2$ generation of his crosses in his 1865 paper, Mendel also determined the ratios expected in later generations. Peas, the species with which Mendel worked, are self-fertilizing plants, and he calculated correctly that under selfing, the proportions of the three genotypes from an $a_1a_1 \times a_2a_2$ hybrid would be $2^N - 1:2:2^N - 1$, in which $N$ is the number of generations of selfing, beginning with the $F_2$. Thus, under selfing, ratios of 3 $a_1a_1$:1$a_1a_2$:3$a_2a_2$ are expected in generation $F_2$, 7:2:7 in $F_3$, 15:2:15 in $F_4$, and 31:2:31 in $F_5$; clearly, the heterozygous class rapidly becomes rare under selfing and ultimately disappears. However, many organisms do not reproduce by selfing, which is clearly the case of species with separate sexes. It was apparent, therefore, that sooner or later the question of the proportions of genotypes expected in various generations would arise in bisexual organisms. Prior to 1900, when Mendel's paper was rediscovered, a biometrical school of genetics had developed that was based on correlation coefficients between parents and offspring, and between other groups of related individuals. This school proposed a "Law of Ancestral Heredity," which held that the two parents each contribute one-half to the total heritage of the offspring, the four grandparents contribute one-fourth each, the eight great-grandparents one-eighth each, and so on. In other words, it was postulated by the biometricians that the preparental individuals, in fact, contribute to the genotypes and phenotypes of later members of any lineage. It was this premise of the biometricians that particularly upset the Mendelians.

Mendel's Law of Segregation (1990) had focused on the purity of the gametes. Thus, according to the Mendelians, a homozygous recessive individual in a segregating population, even if its parents were both heterozygotes, should transmit its own recessive genes, in no way contaminated by dominant parental influences. Hence, preparental ancestral influence could not occur. Some biometricians thought that there was no inconsistency between Mendel's law and their Law of Ancestral Inheritance. This unfortunate error stimulated W.E. Castle to counterattack (Castle 1903). Castle, unaware of Mendel's extension of his calculations to include later generations, pointed out that the deduction of preparental ancestral influence by the biometricians, based on correlation coefficients, was marred by two elementary errors. The first was that the computations of the biometricians had been calculated on the assumption that reproduction is based only on the three-quarters of the population that have dominant phenotypes ($AA + 2Aa$), rather than the entire distribution ($AA + 2Aa + aa$). The second was that the computations of the biometricians had also been erroneous, leading to an increase in the sum of the dominant classes from 75% in $F_2$ to 83.3% in $F_3$ to 85% in $F_4$, ultimately approaching an erroneous limiting value of slightly over 85.355339% in still later generations. These dual errors of the biometricians thus indicated that a permanent effect had been exerted by previous recessive ancestry, which was directly contrary to Mendel's concept of purity of the gametes. Castle also showed that the correct expectations for the summed frequencies of the heterozygous three classes ($AA$, $Aa$, $aa$) in monohybrid segregating populations would not lead to an ultimately limiting value for dominant alleles of slightly higher than 85% in later generations, as some influential biometricians had calculated, but would in fact lead to a gradually increasing value that would ultimately lead to fixation (as Mendel had calculated in 1869). Castle also showed that whatever ratio had been reached during various number of generations of selection, this ratio would remain constant in all future generations once selection was discontinued. Thus, in general, as soon as selection is arrested, the population should remain stable at the allelic frequencies that had been attained at the time selection had been discontinued. Castle had discovered an "equilibrium law" not only for genotypes originally present in equal frequency (1:2:1) in $F_2$, but also for the indefinitely large class of ratios resulting from selection for additional generations for or against one member of a pair of alleles, including fixation of one or the other allele. In 1908 the problem was referred by one of the biometricians to G.H. Hardy, a mathematician, who appropriately regarded it as a very simple problem requiring for solution "only a little arithmetic of a multiplication-table type." Hardy showed that no matter what the frequency of genotypes at the time of arrest of selection, the distribution would remain constant thereafter (Hardy 1908). It was discovered many years later that a German biologist and physician, W. Weinberg, had also in published in 1908 (in fact, prior to Hardy's publication) a solution equivalent to Mendel's and Castle's, except that Weinberg had, for the first time, expressed the results of random mating (panmixia) not in terms of genotypic frequencies, but in the now familiar binomial and more convenient form $p^2AA + 2pqAa + q^2aa$, in which $p^2$ and $q^2$ represent the frequencies of alleles $A$ and $a$. The names of both of these post-Castle discoverers are now attached to this most basic of population genetic expressions, the "Hardy-Weinberg Rule." In 1909 Weinberg generalized the theorem to include multiple alleles at a locus.

## HARDY-WEINBERG RULE

We now illustrate the essential features of the Hardy-Weinberg Rule with three simple numerical examples. First, consider two plants of a hypothetical outbreeding species, one plant homozygous $a_1a_1$ and the other homozygous $a_2a_2$, for adaptively neutral alleles $a_1$ and $a_2$, alleles that produce genotypes that can be classified unambiguously as homozygotes or heterozygotes in all genetic backgrounds and in all environments. All $F_1$ seeds obtained by crossing such $a_1a_1$ plants with $a_2a_2$ plants will be $a_1a_2$ heterozygotes. Suppose individuals from such $a_1a_2$ seeds are allowed to self or to mate at random to produce ultimately large $F_2, F_3, \ldots, F_n$ populations and that these $F_3$ and later-generation populations are grown under identical conditions and allowed to mate strictly at random. Under such conditions it can be shown, "with a little arithmetic of a multiplication-table type," that allelic frequencies will remain $\frac{1}{2}a_1$ and $\frac{1}{2}a_2$, and that genotypic frequencies in the $F_2$ and all subsequent generations will be in the proportions $p^2$ $a_1a_1$: $2pq$ $a_1a_2$: $q^2$ $a_2a_2$. Repeating the calculations, but starting with different proportions of $a_1a_1$ and $a_2a_2$ homozygotes as parents (say 0.9 $a_1a_1$ vs. 0.1 $a_2a_2$), allelic frequencies will be 0.9 $a_2$ and 0.1 $a_2$ in all subsequent generations and genotypic frequencies will be 0.81 $a_1a_1$: 0.18 $a_1a_2$: 0.01 $a_2a_2$ in the $F_2$ and all subsequent generations. Generalizing, the rule is that in large random mating populations with no selection, equilibrium is reached in a single generation at a composition given by expansion of the binomial:

$$(pa_1 + qa_2)^2 = p^2a_1a_1 + 2pq(a_1a_2) + q^2(a_2a_2), \qquad (5\text{-}1)$$

in which p and q are the frequencies of alleles $a_1$ and $a_2$, and $p^2$, $2pq$, and $q^2$ are the frequencies of the $a_1a_1$, $a_1a_2$, and $a_2a_2$ genotypes, respectively. With three alleles of a single locus, the generalized result is obtained by expansion of the trinomial $(pa_1 + qa_2 + ra_3)^2$, by a quadrinomial for four alleles, and so forth. The generalized result for several independently inherited (nonlinked) loci, each with two alleles, is given by expansion of $(p + q)^2 (r + s)^2$. Note, however, that attainment of equilibrium will be slower if the foundation population is made up by mixing genotypes including individuals in which some loci are homozygous and some heterozygous for alternative pairs of alleles—for example, $a_1a_1b_1b_1$ mixed with $a_1a_1b_2b_2$. The slower attainment of equilibrium in such cases results from the fact that movement toward two-locus (or multilocus) equilibrium starts only when each locus becomes heterozygous. Thus in a mixture of $a_1a_1b_1b_1$ and $a_2a_2b_1b_1$ plants, progress toward equilibrium starts for locus a in the $F_2$ generation but will not start for locus b until at least one $b_1b_1$ plant hybridizes with an appropriate non-$b_1b_1$ heterozygous plant. Two additional results become apparent from the preceding exercises. First, if any allele, say allele $a_1$, is dominant over recessive allele $a_2$, the recessive allele $a_2$ may be present in the population at modest frequency (say 0.1), but it may rarely be seen in the population because $q < q^2$. For instance, if recessive allele $a_2$ is present in frequency $q = 0.1$, $q^2 = 0.01$, only one among 100 individuals with genotype $a_2a_2$ is expected in all generations after generation $F_1$. Consequently, large numbers of individuals (~300) would have to be grown per generation to assure (p ~ 0.95) the appearance of at least one $a_2a_2$ individual in each generation.[1]

---

[1] As a rule of thumb, given that the probability of an infrequent desired event is p, the number of individuals to be sampled to give a 95% chance of at least one occurrence of this event will be ~3/p. Thus, if the probability of an expected event is $1/64 = 0.0156$, it can be shown that the precise number of individuals to be sampled is 190.3, whereas 3/p = 192, a closely similar value.

Simple Hardy-Weinberg equilibrium of the sort postulated here is unlikely to prevail exactly in any real population, wild or cultivated, because attainment of equilibrium will likely be upset by one or more of the following: departures from random mating, differences in the fitness of different genotypes, linkage, mutation, or random drift.

## RANDOM MATING

Strictly speaking, the term *random mating* should be applied only to situations that satisfy two criteria: (1) every member of a population has an equal chance to produce offspring and (2) any female gamete is equally likely to be fertilized by any male gamete. It can be questioned whether the theoretical form of random mating is ever fulfilled exactly, because some form of selection, natural or human, is likely to intervene at some stage or stages in the life cycle of cultivated plants, thus violating the first criterion (equal chance to produce offspring). For example, in cultivation breeders may often unknowingly favor particular plants, thus inadvertently introducing selection into the system. It is also unlikely that the second criterion is ever fulfilled exactly. Environmentally induced differences in time of pollen shed and in megagametophyte maturity, the vicissitudinous chances of location within natural populations as well as in cultivated nurseries in respect to prevailing wind direction, and the like, make it improbable that fertilizations are ever entirely random events, either in nature or in cultivation. If, however, all conditions necessary for the theoretical form of random mating are met, three predictions can be made concerning the genetic composition in later generations of large outbreeding populations: (1) allelic frequencies will remain nearly constant over generations, (2) levels of heterozygosity and homozygosity will also remain nearly constant, in accord with Hardy-Weinberg proportions, and (3) genetic relationships between individuals within populations in Hardy-Weinberg equilibrium will soon reach $F \sim 0$ and remain nearly constant over generations at $F \sim 0$.

## DEPARTURE FROM RANDOM MATING

We first consider four of perhaps the most important kinds of departures from random mating that breeders can impose on their populations. These departures are based on the two kinds of the mating of like-to-like, or on the two kinds of the mating of unlike individuals that are possible in populations, namely, (1) genetic assortative mating, (2) genetic disassortative mating, (3) phenotypic assortative mating, and (4) phenotypic disassortative mating. The criterion of likeness or unlikeness is either relationship (ancestry) or appearance (phenotypic resemblance or phenotypic difference). Genetic assortative mating (mating of relatives) is the most powerful of all mating systems for assembling favorable combinations of genes, because determination of genetic relationships does not depend on the ability of the breeder to identify similarities or differences in allelic composition, or on ability to identify similarities or differences in phenotypic appearance or performance, but only on ability to keep accurate pedigree records. Genetic assortative mating is usually referred to as *inbreeding* in genetics and in plant or animal breeding. The primary effect of genetic assortative mating (inbreeding) is to increase the probability that offspring will inherit the same alleles from both their male and female parents. This tends to lower heterozygosity, leading ultimately to possible fixation of identical alleles and, hence, to possible fixation of the phenotype to the extent it is under genetic control. In other words, genetic assortative mating is not affected by mistaking the effects of environment for the effects of genes, or by the difficulties that can be caused by variations in gene expression owing to such factors as dominance or shifts in the genetic background. Inbreeding is thus especially effective in fixing genes that govern characters of low heritability, the characters that are nearly always the most difficult to deal with in plant breeding.

Intense inbreeding also leads to ever-increasing numbers of noninterbreeding families, many of which are likely to suffer from so-called inbreeding depression. Such families often die out (become extinct) on their own, or poorer families are appropriately discarded to keep total population numbers within practical limits. Thus, in actual practice, intense inbreeding, without selection, is virtually impossible. The increase in genetic variability that occurs will obviously be mostly between or among lines (families), because as homozygosity increases, genetic variability within families rapidly decreases toward zero with the more intense forms of inbreeding. At the same time, among-line (among-family) differences increase as different allelic combinations become fixed in different families. The important practical consequences of high genetic correlations within families and wide differences among families are that genetically different inbred lines, when hybridized, sometimes complement one another genetically and "nick" to produce superior monogenotypic $F_1$ hybrid progeny that are also highly uniform phenotypically, whereas genetically similar lines tend to produce inferior hybrids. $F_1$ hybrid varieties have become increasingly important in plant breeding, especially during the last half of the twentieth century, regardless of the original mating system of the species.

The main effect of genetic assortative mating is to increase homozygosity as a result of inbreeding. The most useful quantitative measure of inbreeding is the coefficient of inbreeding (F), devised by Wright (1921). The coefficient of inbreeding, which is based on the number and closeness of ancestral connections between male and female parents, takes the form

$$F_x = \Sigma_A (\tfrac{1}{2})^{N+N+1}(1 + F_A), \qquad (5\text{-}2)$$

in which *A* is an ancestor common in the pedigree of both the male and the female parents of individual *A*, *N* is the number of generations between both the male parent ancestor and *A*, and $N_2$ is the number of generations between the female parents and ancestor *A*. With self-fertilization (the mating scheme most common in plant breeding) $N_1 = N_2 = 0$, so that Equation 5-3 reduces to

$$F = \tfrac{1}{2}(1 + F'), \qquad (5\text{-}3)$$

in which prime indicates the inbreeding coefficient of the preceding generation. The coefficient of inbreeding takes value zero in random mating populations (a usual starting place in outcrossing plants), and F approaches 1.0 within a few generations as selfing proceeds. Under selfing, development of genetic correlations among individuals within families follow the series 0, 0.5, 0.75, 0.875, . . . , 1; complete homozygosity will be the case result at any locus when F = 1. Selfing is therefore particularly effective in increasing all genetic correlations within families toward unity. Homozygotes are highly prepotent because they produce only one kind of gamete; multiple heterozygotes are much less prepotent because they produce many different kinds of gametes and, hence, many different kinds of progeny. Selfing is, in addition, the most effective mating system for increasing genetic divergence among different families. Thus, when homozygous lines are developed by selfing genetically variable but unrelated source populations, the inbred lines that emerge from different source populations are, owing to segregation, likely to carry differing arrays of alleles of many unlinked loci. Genetic assortative mating (selfing in particular) practiced in different source populations is therefore especially useful in quickly assembling and stabilizing, within different homozygous inbred families, differing genotypes among which only a very few on intercrossing are likely to produce monogenotypic hybrids that are remarkably productive. Thus inbreeding in different source populations has been incorporated into many modern breeding schemes, and it is especially effective in dealing with traits of low heritability.

As seen in the preceding paragraphs, the term *genetic assortative mating* can be applied whenever the mating individuals are more closely related by ancestry than when matings are at random. However, in plant breeding the term *inbreeding* is usually applied only when matings are made between closely related individuals, and particularly when mating is by selfing. The second type of departure from random mating based on relationship (ancestry) calls for mating of unrelated individuals; this system is commonly called *outbreeding.* The third type of departure from random mating is genetic disassortative mating, which is rarely practiced within closed populations of the type we have been considering. Its major use in plant breeding is in connection with crossing strains with different origins (unrelated strains) for the purpose of broadening the genetic base in germplasm conservation projects. Such phenotypic disassortative mating is useful in maintaining genetic diversity in populations that serve as sources of alleles; mating unrelated types counteracts random loss (erosion) of genetic diversity and helps to preserve potentially useful allelic diversity. Phenotypic disassortative mating sometimes also finds applications in closed breeding populations; choosing contrasting parents may be useful in correcting opposing weaknesses. Mating of unlikes tends to reduce genetic correlations among relatives. Disassortative mating is perhaps the most conservative of mating systems and is the mating system that best holds populations together and best prevents loss of genetic variability.

An idea advanced earlier in this chapter was that simple Hardy-Weinberg equilibrium is not likely to prevail exactly in any real population, wild or cultivated; this is because attainment of equilibrium is likely to be upset by one or more factors, other than mating system, that can lead to departures from random mating, including selection, linkage, mutation, or random drift. We now consider the effects of each of these factors.

## Fitness or Relative Performance

Fitness is conventionally defined in terms of relative survivability or relative performance: the more progeny left by a genotype or the more progeny any genotype produces relative to other genotypes, the more fit it is. Fitness, often symbolized by $w$, is conventionally defined by reference to some chosen genotype, often a genotype arbitrarily designated to be maximally fit or maximally productive.

$$a_1 a_1 \qquad a_1 a_2 \qquad a_2 a_2$$

$$w = 1 = 1 - S\, a_1 a_2 = 1 - S\, a_2 a_2$$

If genotype $a_1 a_1$ leaves 100 progeny, genotype $a_1 a_2$ leaves 90 progeny, and genotype $a_2 a_2$ leaves 80 progeny per generation, the coefficients of fitness are $Sa_1 a_2 = 0.10$ and $Sa_2 a_2 = 0.20$. Thus, by this convention smaller values of S imply higher fitness, as follows: S = 0 (most fit) to S = 1 (lethal).

Many mathematical representations based on coefficients of fitness or productivity such as these have been developed to predict the changes that may be expected to occur with selection in various kinds of populations. However, it is widely accepted that few phenotypic effects having to do with fitness or productive capacity in higher plants are, in fact, related in simple ways to only one or a few genetic loci. Although some phenotypic expressions may be initiated by only one or a few genetic loci, final phenotypic expressions often appear to be subject to significant modifications resulting from interactions with alleles of several to many other loci and/or to differing interactions with the environment. Reliable information concerning the attributes of fitness is probably always inadequate to at least some degree, and, hence, it has nearly always been necessary to reduce mathematical representations of the evolution of fitness or productivity to mere skeletons of their probable real complexity. Consequently, only the most simple and straightforward single-locus cases are introduced at this point, and consideration of more complex multilocus situations are delayed until the effects of linkage, epistasis, and close inbreeding have been discussed in more detail in later chapters.

In outbreeding populations any plainly unfit or plainly unproductive dominant alleles are expected to be eliminated rapidly, because they will be exposed to selection in all generations. In contrast, any plainly unfit recessive alleles (*aa*) have the advantage of being able to hide in heterozygotes (*Aa*), which are relatively frequent ($2pq$) in heavy outcrossers. Consequently, unfit recessive alleles are expected to take the long road to elimination in outbreeding populations. Unfit alleles of intermediate dominance are expected to be reduced in frequency more rapidly than equally unfit full recessives, because the intermediate dominance of the former exposes them to selection in heterozygotes. All of the aforementioned expectations for single loci are, for the most part, realized in largely outbreeding populations. Moreover, fitness values may vary widely from generation to generation with fluctuations in the environment and/or changes in background genotype, and the amount of outcrossing may differ from generation to generation. The result is that frequency changes in outcrossers often deviate widely from simple expectations.

Heavily inbreeding populations, especially naturally self-pollinating populations, are expected to, and in fact often do, move rapidly toward elimination of plainly unfit alleles and toward fixation of fit single-locus homozygotes. This is because heavy inbreeding rapidly leads to large values of F that quickly expose all alleles to selection (especially full recessives), with the result that the less fit alleles are quickly eliminated; contrariwise, moderately fit alleles are expected to take a slower road to extinction, and very highly fit alleles are soon expected to reach high frequency. The population behavior of less intense inbreeders (i.e., species in which the mean outcrossing rate is on the order of 5–10%) generally tends to be somewhat closer to that of heavy inbreeders than to that of outbreeders.

## Linkage

Loci located on different chromosomes segregate independently, except for the case of possible interlocus interactions. This is also the case for loci located sufficiently distant from one another on different arms of a chromosome or on different chromosomes. Random mating populations that start from multiple heterozygotes are expected to reach equilibrium for unlinked loci in a single generation, but move slowly toward equilibrium from other starting points. Thus loci that are linked will ultimately come to joint equilibrium, but at a slower rate determined by selective values and the closeness of the linkage. However, even loose linkage often delays the approach to equilibrium beyond the normal short duration of many (or most) plant-breeding programs. In the extreme case of no recombination, all completely linked loci behave as a single functional locus so that no movement occurs toward joint equilibrium. As noted earlier, genetic assortative mating (inbreeding) will, by promoting homozygosity, affect allelic interchanges. Furthermore, very close inbreeding, in itself or in combination with even modest linkage, can reduce effective recombination to very low levels, causing all loci to behave as if they were very tightly linked. These phenomena have such important implications concerning pseudo-overdominance that it is appropriate to postpone discussion of them until details are discussed later in this chapter in terms of parameters measuring recombination rates.

## Mutation

Mutation rates are usually low, approximately on the order of 1 mutant per 100,000 to 1 per 1,000,000/locus/generation. Consequently, mutation can usually be neglected in breeders' nurseries in which numbers of individuals are usually small. However, the importance of mutation in long-term evolutionary change in large wild populations is obvious. Many useful mutants have appeared and been identified, especially in large commercial plantings of cultivated plants, and they are sometimes perpetuated by alert farmers and/or seedsmen. Thus it is worthwhile for plant breeders to remain watchful for potentially useful variants in their breeding populations and, more particularly, in their varieties after they are released to large-scale commercial production.

## Random Mating in Small Populations

When population sizes are small, as is often the case in practical plant-breeding nurseries, two additional factors, chance and inadvertent inbreeding, must be taken into account in predicting population change. The role of chance can be illustrated by considering two hypothetical diploid populations, one made up of 10 individuals per generation (a common size in plant breeding) and the other made up of 1,000 individuals per generation. Consider a locus with two alleles, $a_1$ and $a_2$ present in equal frequencies, $p = q = \frac{1}{2}$, in both the smaller and the larger population. The small population and larger population thus arise each generation from a sample of 20 and 2,000 gametes, respectively. The total numbers of $a_1 + a_2$ alleles in the smaller and larger populations and their standard deviation, respectively, are

$$10 \pm (10 \times 10/20)^{\frac{1}{2}} = 10 \pm (5)^{\frac{1}{2}} = 10 \pm 2.2361$$

$$\text{or } 1000 \pm (1,000 \times 1,000/2,000)^{\frac{1}{2}} = 1,000 \pm (500)^{\frac{1}{2}} = 1,000 \pm 22.36$$

In the small population the standard deviation is a relatively large percentage ($\sim$ 11%) of the gamete number, but a much smaller percentage of the gamete number in the larger population ($\sim$ 1%). Hence, the proportions (p and q) of alleles $a_1$ and $a_2$ are expected to fluctuate widely from generation to generation in the smaller population, but are expected to remain relatively constant in the larger population. These random fluctuations in p and q arising from accidents of sampling are called *genetic drift*. Allelic frequencies ultimately become either 1 or 0 in both populations after a sufficient number of generations even though random mating has been the practice in both populations, but allelic frequencies clearly fluctuate much more rapidly in smaller than in larger populations. Thus random fluctuations in allelic frequencies can cause random genetic drift, especially in small populations, because allelic frequencies are likely to "drift" up or down rapidly as a result of chance alone. Moreover, allelic frequencies are likely to become 0 or 1 within a few generations in very small populations; thus, even quite favorable alleles can often be lost solely by accident in small populations, whereas even rather unfavorable alleles can become fixed, also quite by accident. Drift is beyond the control of the breeder, except by keeping population sizes large enough to hold random fixation to low levels.

Another major factor affecting the genetic composition of small populations is that inbreeding leads to reduction in the number of individuals contributing offspring. In any small population, more consanguineous matings will occur than in larger populations. The decrease in heterozygosity as a result of the genetic assortative mating will be $\frac{1}{2}N$ per generation, in which $N$ is the number of monoecious individuals whose gametes unite completely at random, including self-fertilizations. In a random mating population of 10 monoecious diploid individuals in which 100 loci are heterozygous, $\frac{1}{2}N$ or ~5% (or ~5 loci) are expected to become fixed in the first generation, and in each subsequent generation additional heterozygous loci are expected to reach fixation. When the numbers of female and male parents differ, the effect is roughly equivalent to reducing the effective breeding size of the population to nearer that of the less-frequent parent. This reduction can be estimated by

$$\tilde{N} = 4N_f N_m / N_f N_m \qquad (5\text{-}5)$$

in which $N_f$ and $N_m$ are the numbers of female and male parents and $\tilde{N}$ is the effective population size. Suppose the next generation is produced by only 5 seed parents and only 50 pollen parents. Heterozygosity is expected to be reduced at the same rate as if there were only a total of $\tilde{N} \sim 18$ breeding individuals in the population. Effective population size clearly depends far more on the less frequent than the more frequent sex. With a limited number of seed parents but a very much larger number of pollen parents, as is often the case in plant breeding, $\tilde{N}$ may often be only slightly larger than $N_f$. Thus loss of genetic variability owing to sampling accidents can be substantial in small plant-breeding nurseries or if few female (or male) parents leave offspring. Nevertheless, the rate of decay in heterozygosity in random mating populations ($\frac{1}{2}N$ per generation) is relatively slow even when population sizes are quite small. Thus, if the goal is to increase homozygosity, assortative mating (genetic or phenotypic) systems are more effective than other systems. The utility of random mating schemes in plant breeding is therefore largely limited to preserving desirable alleles in the gene bank or in breeders' collections, or in recurrent selection programs. Nevertheless, if the goal is to preserve desirable alleles in populations, some disassortative mating system is likely to be much more cost-effective than a scheme involving random mating, unless disassortative mating has a much higher labor requirement than simply allowing plants to mate at random.

## EXAMPLES BASED ON HYPOTHETICAL POPULATIONS

We now consider the expected course of genetic changes in two hypothetical populations, one modeled on corn, a heavy outcrosser in which $\sim 90\%$ of matings are more or less random outcrosses and $\sim 10\%$ are self-pollinations, and the other modeled on barley, a heavy inbreeder in which $\sim 98\%$ to $\sim 99\%$ of matings are self-pollinations and $\sim 1\%$ to $\sim 2\%$ are more or less random outcrosses (Chapter 4). To simplify, we assume that both hypothetical populations were started by crossing a pair of fully homozygous inbred lines and that the corn $F_2$ was produced by nearly complete natural random mating among $F_1$ hybrid plants, whereas the barley $F_2$ was produced by nearly complete natural selfing of $F_1$ hybrid plants. We also assume that population sizes are very large in $F_2$ and in all subsequent generations. Corn has 10 chromosomes (20 chromosome arms), and barley 7 chromosomes (14 chromosome arms). Our computations, unless otherwise specified, will be based on 10 loci, each with two selectively neutral alleles (no mutation), and it will be further assumed that all loci are located sufficiently far apart that physical linkage will not be involved. Under such assumptions, all allelic frequencies are expected to be $\frac{1}{2}$ in $F_1, F_2$, and all subsequent generations in both the hypothetical random mating (corn) and the hypothetical selfing (barley) populations.

The corn population is expected to attain equilibrium in $F_2$ and to retain the same average allelic composition ($p = q = \frac{1}{2}$) and genotypic composition $p^2 a_1 a_1 : 2pq(a_1 a_2) : q^2(a_2 a_2)$ per locus in all generations. With $N = 10$ loci and 2 alleles/locus, $3^N = 59,049$ different 10-locus genotypes are expected in $F_2$ and all subsequent generations. Among these 59,049 genotypes only one genotype is expected to be heterozygous at all 10 loci, whereas $2^N = 1,024$ genotypes are expected to be homozygous at all 10 loci, and $3^N - 2^N - 1 = 58,024$ genotypes are expected to be heterozygous at some loci but homozygous at other loci. The smallest complete $F_2$ population (at least one individual of each type present) is expected to be $4^N = 1,048,576$ individuals. If a particular homozygote were desired and could be identified, its expected frequency would be $(\frac{1}{4})^{10}$ or $\sim 9.5 \times 10^{-7}$, so that vast numbers of plants would have to be examined to allow even a small chance of detecting such a genotype. In any real situation the number of alleles would almost certainly be larger than 10; it is also likely that there would be more than 2 alleles at many loci. Assuming 20 loci (1 per chromosome arm in corn), each with 2 alleles, about $3.5 \times 10^{10}$ different 20-locus genotypes are expected in the corn $F_2$ and in all subsequent generations. Complexity would also increase rapidly if there were more than 2 alleles per locus. Assuming 3 alleles at each of 20 loci, more than $3.6 \times 10^{15}$ 20-locus genotypes are expected to be present in each generation of the outbreeding population. The smallest complete population would clearly have to be very large in each of the aforementioned cases, and the expected frequency of any given 10 (or 20)–locus genotype would be very small. Consequently, impractically large numbers of plants would have to be examined to provide a reasonable chance of detecting any desired genotype, even if unrealistically small numbers of loci and alleles are assumed. Thus, all or very nearly all real random mating populations will be incomprehensibly diverse genetically.

Both the selfing and the outbreeding populations are expected to have the same genetic composition in $F_2$. The selfing population is, however, not in equilibrium, and inbreeding is expected to decrease the complexity of the population very rapidly in the first several generations. The parents of the hypothetical populations were postulated to be highly inbred lines. Consequently, it is quite unlikely that any conspicuously unfit alleles would have been introduced into either the outbreeding or inbreeding population. However, in the $F_2$ or any later generation occasional deleterious alleles might have been introduced into both populations by mutation. Such deleterious alleles (especially dominant alleles) would quickly be exposed to selection and, hence, be subject to early elimination from the inbreeding population; however, deleterious recessives might survive for some generations in the outbreeding populations by hiding in heterozygotes. Rare truly overdominant mutant alleles (heterozygotes superior to both homozygotes of the same locus) might conceivably be encountered. Such alleles might conceivably lead to excess heterozygosity for a few generations in the outbreeding population but would be unlikely to have a detectable effect on heterozygosity in the inbreeding population. Drift might also affect both populations, but its effects would be small, provided population sizes were on the order of a few hundreds of individuals per generation. No linkages were assumed among any loci, but should unknown linkages exist, they would delay attainment of equilibrium in both kinds of population. However, the $F_1$ is heterozygous at all $N$ loci in the two hypothetical kinds of $F_1$ populations, and in the $F_2$ half of the loci are expected to remain heterozygous. Consequently, any loose linkages would likely become victims of segregation in those generations in both kinds of population. Furthermore, in the outbreeding population, in which half of the loci are expected to remain heterozygous in $F_3$ and all subsequent generations, only tightly linked pairs of loci would be expected to remain in disequilibrium for more than a very few generations. Attainment of equilibrium in the inbreeder, however, differs dramatically. The $F_1$ generation of the selfing population is expected to be heterozygous at all $N$ loci, in the $F_2$ at $1/2$ of its loci, and the same rate of reduction in heterozygosity is expected to continue over generations according to the series 1, 1/2, 1/4, 1/8, 1/16, 1/32, 1/64, 1/128. . . . As heterozygosity falls, the selfing population is expected to approach an ultimate equilibrium state of $2^N$ homozygous lines in equal frequencies according to the series 0, 0.5, 0.25, 0.125, 0.0625, 0.0313, 0.0156, 0.0078. . . . Thus average heterozygosity is expected to fall to less than 1% and homozygosity to increase to more than 99% by $F_8$. With 10 loci, each with two alleles, the barley population should rapidly become much less complex. By generation $F_8$, assuming no differential selection, it is expected to contain only about 1,000 different complete homozygotes together with about 24 near-homozygotes still segregating for perhaps one or two loci. Fewer homozygotes and near-homozygotes are expected if some are reproductively inferior.

We turn now to the effects of disturbing factors that are likely to play a large role in any real population. As shown previously, about 93% of the recombination within lines in the selfing population will have occurred by generation $F_5$. Very little additional recombination is expected thereafter, so most of the linkages between loci that survive within nearly homozygous lines are likely to continue to survive until such lines reach full homozygosity. At this point the occasional outcrosses ($\sim 1$ to $\sim 2\%$ per generation in barley) that occur between different homozygous or very near homozygous genotypes might well enter the picture and assume considerable importance in generating novel genotypes. The $F_1$ hybrids resulting from such outcrosses are likely to be heterozygous at many loci, with the result that each individual $F_1$ plant resulting from outcrossing in generations $F_2$ or later might be expected to initiate a burst of segregation that could possibly culminate ultimately in $2^N$ different near-homozygous multilocus genotypes. Theory (review in Turner 1967) holds that when crossover values are small relative to the intensity of selection, stable nonrandom associations of alleles are expected to develop and persist in populations. Provided linkage is sufficiently tight and selection sufficiently strong, genetic variability is expected to become so organized that correlations between loci will be complete and $D$, the linkage disequilibrium parameter, will take its maximum value. Theory also indicates that any factor that restricts recombination will have an effect similar to that of linkage in binding concordant alleles together and that positive assortative mating in particular can lead to sharp restriction in recombination (Jain and Allard 1966). The way this works can be illustrated by two individuals with genotypes $a_1a_1b_1b_1$ and $a_2a_2b_2b_2$ (loci $a$ and $b$ unlinked) in the fourth or later generations of predominantly selfing populations. Because of predominant selfing, these individuals will nearly always produce only $a_1a_1b_1b_1$ and $a_2a_2b_2b_2$ progeny generation after generation, so that the $a_1b_1$ and $a_2b_2$ alleles are expected thereafter to remain correlated within all such lineages just as if they were completely linked. But when sporadic hybridizations occur between the two lineages, leading to $a_1a_2b_1b_2$ heterozygotes, a burst of recombination and segregation is expected, especially during the first few generations, during which period $a_1b_2$ and $a_2b_1$ gametic types may be produced and the associations may be broken. When such intercrosses between different homozygous lines occur only once in 50 generations (2% outcrossing) or so, as expected in the case of many predominantly selfing plant species such as barley, heterozygotes will clearly be infrequent and, hence, recombination is expected to be severely restricted because effective crossing over occurs only in heterozygotes. This also applies to multilocus situations. Assume four unlinked diallelic loci and epistasis such that the gametic combinations $a_1b_1c_1d_1$ and $a_2b_2c_2d_2$ yield homozygotes and a heterozygote that are superior in fitness and that the other 14 gametic combinations are adaptively inferior. During the course of any generation, selection will favor the $a_1b_1c_1d_1$ and $a_2b_2c_2d_2$ combinations, causing their frequencies to increase from early to later stages in the life cycle relative to the 14 other recombinant gametic combinations. Thus "supergenes" will soon form (Turner 1967) such that only 2 of the 81 possible genotypes will occur in the population. Accordingly, correlations between all loci may become effectively complete, $D$ may take its maximum value of unity, and the entire genome may ultimately congeal into only two supergenes.

It would be helpful to have a quantitative measure of the joint restriction of recombination caused by linkage, and of that caused by the mating system. For present purposes, the rate at which $D$ (the linkage disequilibrium parameter) converges to 0 for neutral alleles is convenient. For two loci (Weir and Cockerham 1973) this rate is given by

$$1 - \tfrac{1}{2}\{1 + \lambda + s/2 + [(1 + \lambda + s/2)^2 - 2s\lambda]^{\frac{1}{2}}\} \qquad (5\text{-}4)$$

in which $\lambda$ is the amount of linkage ($0 \geq \lambda \geq 1$) and $s$ is the probability that an individual chosen at random in any generation is the offspring of a single individual in the previous generation; $t = 1 - s$ is the probability that it has two parents. Note that $\lambda$ and $s$ enter this expression in the same way, so that the magnitude of their individual effects on the rate of decay of $D$ is equal. With random mating ($s = 0$ as in the corn population) and no linkage ($\lambda = 0$), the crossover value $c$ is 0.50, so that $\tfrac{1}{2}$ of any disequilibrium is lost in the next mating cycle. With random mating $s = 0$, $t = 1$, but with tight linkage $\lambda = 0.98$, $c \sim 0.01$; hence, only about 1% of any disequilibrium is lost per generation. The asymptotic approach of $D$ to 0 for neutral alleles under random mating is the geometric rate $1 - c$ per generation, so that $D$ in any generation, $t$, is given by $D_{(1)} = (1 - c)^t D_{(0)}$. With no linkage but 98% self-fertilization, the rate of decay in $D$ is also ~1% per generation, or ~$\tfrac{1}{50}$ as large as with random mating in the absence of linkage. When tight linkage is combined with heavy selfing, recombination is reduced to a point at which very little selfing is required to hold favorable combinations of alleles together. For example, if $\lambda = 0.98$ and $s = 0.98$, decrease in $D$ (loss of disequilibrium per generation) is given by $D^t = (1 - 0.0004)^t = $ only 0.9992—that is, loss of existing disequilibrium is << 1% per generation.

Obviously, there is an important difference in the effects of restriction of recombination as a result of linkage and restriction of recombination as a result of selfing. Linkage protects favorable combinations of alleles, located physically close to one another on the same chromosome, from breaking up, but it does not protect concordant allelic combinations for loci located on different chromosomes from breakup. Inbreeding, in contrast, restricts recombination between all loci, whether they are located on the same or different chromosomes. Theory therefore predicts that heavy inbreeding alone is capable of binding the entire genotype together and, hence, that it can be an exceptionally effective mechanism in organizing the entire gene pool (especially favorably interacting alleles of different loci) of heavily inbreeding populations into integrated systems.

## SYNOPSIS

The simplest population genetic assumptions applied to hypothetical populations originally heterozygous at $N$ diallelic loci lead to the expectation that heavily random mating outbreeders should approach Hardy-Weinberg equilibrium in a single generation and that they should thereafter remain heterozygous with all loci at 50% heterozygosity. Hypothetical heavily selfing populations, in sharp contrast, are expected to approach equilibria featuring $2^N$ homozygous or very nearly homozygous multilocus genotypes within about six generations or so. Thus, with 10 loci $2^{10} = 1,024$, homozygous multilocus genotypes are expected by the time near homozygosity is achieved.

Experimental data from studies of actual predominantly outcrossing populations, both wild and synthesized from cultivated materials, show that such populations do, in fact, conform closely to the preceding expectations. All outbreeding populations are, in fact, heterozygous at many loci and are intolerant of inbreeding. Yet allelic frequencies and multilocus genotypic frequencies change over time, most likely owing to viability differences among different genotypes, and outbreeding populations, in fact, often carry somewhat greater numbers of deleterious alleles (e.g., chlorophyll deficients) than expected (Sprague and Schuler 1961).

Observed departures from simple expectations are, however, usually dramatically greater in actual heavily selfing populations than in random mating populations. The most striking departure is that the numbers of surviving homozygous lines actually observed in heavily selfing populations have consistently been very much smaller than the $2^N$ expected. For example, in populations synthesized from crosses between homozygous cultivated parents differing at, say, 10 or even many more electrophoretically distinguishable loci, the numbers of near-homozygous or fully homozygous 10-locus genotypes that actually survived into generation $F_6$ were, in fact, always fewer than 25 and often fewer than 5; both of these observed numbers of survivors are far fewer than the $2^N = 1,024$ expected. This suggests that most of the hundreds (or thousands) of 10-locus homozygotes, or near homozygotes, present in early generations were unfit to some degree and that this was almost certainly the primary reason for their rapid elimination from the population. It also suggests that the occasional more fit survivors rapidly became ascendant because of their adaptive superiority. It should be emphasized that population sizes in all generations of the experimental populations were sufficiently large to reduce the probable effects of drift to insignificance or near insignificance. This conclusion, reached from studies of many different experimental populations, is supported by studies of electrophoretically detectable variants in many large long-established wild populations of heavy selfers. Such studies show that the surviving multilocus genotypes in such populations are often adaptively superior; typically, genotypic variability in large heavily selfing populations has been found to vary from very small (only 1 to 3 multilocus genotypes present per population) to moderate (at most 10 to 25 multilocus genotypes present) when $N$ (the number of alleles) was greater 30, so that many more than $1 \times 10^7$ multilocus genotypes would be expected in such populations if viability differences were small. Thus, various kinds of evidence make clear that mating systems featuring heavy selfing are favorable for rapid development and long-term maintenance of patchwork patterns of epistatic interlocus combinations of alleles that promote high adaptedness within fine-grained local environments. We examine the development of such multilocus and multiallelic systems in some detail in later chapters, especially in Chapters 10 and 11.

It is also important to note that artificial imposition of even low levels of outcrossing (say ~5%) on experimental populations of predominant inbreeders leads to hybridization, segregation, and recombination, which rapidly break up favorable multilocus combinations of alleles. Reductions in adaptedness have consistently been substantial whenever levels of outcrossing have been increased artificially to 5% or more in predominantly selfing cultivated populations. Analogously, the population structures of species that outcross at rates of 5% or more (e.g., sorghum) are usually quite complex, often featuring many multilocus genotypes, including only a few that are homozygous at most loci. In general, however, such species are somewhat tolerant of inbreeding, and plant breeders have developed many highly homozygous and reasonably vigorous inbreds that, when crossed, produce very high-yielding single-cross hybrids.

Clearly, the general pattern of the plant-breeding program appropriate to any given cultivated species depends, in large part, on its reproductive system. We now develop this concept further in terms of the kinds of population structure that have developed in cultivated species as various reproductive systems have been imposed on them subsequent to their domestication. More detailed descriptions of appropriate breeding plans will be given in Parts II and III.

## PLANT-BREEDING SCHEMES: HISTORICAL

In carrying out the original domestications of wild plants it seems reasonable to assume, lacking direct evidence other than the variability observed in present-day wild populations, that the earliest domesticators did what would have come naturally—that is, they would have saved seeds (or other propagules) from an adequate number of "desirable" wild or near-wild individuals to meet planting requirements in their cultivated fields. As noted earlier, all or virtually all populations of wild plants, regardless of reproductive system (inbreeding or outbreeding), are genetically variable to a greater (outbreeders) or lesser (inbreeders) extent. Consequently, it is likely that the earliest cultivated populations all contained several to many different multilocus genotypes and that the practice of "mass selection," that is, choosing many "desirable" individuals as parents of the next generation, continued over the centuries. Certainly, wild populations and present-day landraces of predominantly selfing species are usually made up of a number of different genotypes, and open-pollinated populations of outcrossers are much more conspicuously variable genetically. The first written accounts of the extent of diversity within cultivated populations of inbreeders came in the eighteenth and nineteenth centuries with methodical attempts at plant breeding. Van Mons (Belgium), Knight (England), and Cooper (United States) all wrote of improvements achieved by selecting superior types from their landraces. More informative work followed in the nineteenth century. John LeCouteur, a farmer of the Isle of Jersey, noted the diversity of types in his wheat fields. He also noted that there were differences in the agricultural value of different progenies and that all of the offspring of most single plants were remarkably alike. LeCouteur, who published a summary of his work in 1843, apparently was the first to set down clearly the value of selecting individual plants in the improvement of self-pollinating small grains, although Patrick Sheriff (Scotland) had earlier used the same methods in breeding some extensively grown varieties of wheat and oats. In the mid-eighteenth century, F. Hallett (England) practiced single-plant selection with predominantly self-pollinating small grains. In common with many breeders of the time, he believed in the inheritance of acquired characters and he consequently grew his progenies under the best of conditions, continuing to select the best plants in each generation. In failing to observe that continued selection in progenies of autogamous species derived from single plants is largely ineffective, he contributed less to understanding of the structure of such populations than his predecessors. His false premise (inheritance of acquired characters) did not, however, prevent him from isolating a number of valuable varieties from variable landraces, the most noteworthy being Chevalier barley.

By far the most notable nineteenth-century contribution to understanding the genetic structure of cultivated populations was due to Louis de Vilmorin (1856) of the Vilmorin seed firm of France. He developed a method of line selection featuring progeny testing in each generation. When he applied this method to wheat, a highly selfing species, selecting the best plants in each generation for many generations, he detected few if any changes in appearance or performance. But selection within progeny lines for high sugar content in sugar beets, an outcrossing species, during the period 1850 to 1862, led to substantial inbreeding and noticeable inbreeding depression. Yet intense selection for vigor and, in particular, for higher sugar levels so markedly improved the sugar content of the beets that the sugar beet industry became economically viable. With this work the differences in population structure and the differing effects of selection in self- versus cross-pollinated species had been discovered. However, it remained for the geneticists and plant breeders of the early twentieth century, notably Johannsen, Nilsson-Ehle, and East, to explain the genetic basis of pure lines and the genetic differences in the population structure of inbreeding versus outbreeding species.

In predominantly selfing species, such as barley and wheat, all loci are quickly forced to homozygosity by the natural mating system, and each local population soon becomes a mixture of different homozygous or very nearly homozygous lines. However, the less fit homozygous lines are unlikely to survive, as a result of selection, with the result that within-population genetic variability is soon reduced to much below original levels, especially in populations descended from a single individual. Although occasional outcrosses occur among plants of each local population, many of the outcrosses that occur clearly must be between individuals of the same, or at least closely similar, genotypes. Hence, relatively little segregation occurs in the progeny of such crosses, with the result that long-established inbreeding populations change relatively little genetically from generation to generation. Outcrosses also occur occasionally between different genotypes in the same local population, but less frequently with migrants from other local populations. Although most segregants from outcrosses between different genotypes are inferior and hence do not survive, occasional outcrosses sometimes segregate to produce superior homozygous descendants, some of which may be incorporated into the population. Thus the adaptedness of heavily selfing cultivated populations almost certainly improved gradually over the centuries as the frequencies of novel superior homozygous segregants increased owing to natural and/or human-directed "mass selection." Pure-line breeding became common in northern temperate zones in the latter part of the nineteenth century, and during the next 50 years or so farmers and seedsmen isolated many distinct true-breeding monogenotypic lines often from single individuals, selected from their "landrace varieties" that had been developed by single-plant selection. Evidently, many of the new, more uniform "pure-line" varieties were superior to the landraces from which they had been isolated. At any rate, quite uniform varieties almost entirely replaced their more variable predecessors in the latter decades of the nineteenth century and the first few decades of the twentieth century. Highly variable landraces persist to this day, especially in subsistence agricultures, for reasons to be considered in Chapter 17. In the first few decades of the twentieth century both mass selection and the isolation of pure lines from landraces and "farmers" varieties of self-pollinated species were almost entirely abandoned in advanced agricultures. Instead of waiting for occasional fortuitous outcrosses to occur in traditional mass selected populations, seedsmen and members of the newly founded profession of plant breeding adopted deliberate artificial hybridization between carefully selected parents as a means of combining in a single segregant line desirable alleles found separately in the parents. The segregating generations were handled in a variety of ways, often by the so-called pedigree, bulk population, single-seed-descent, or backcross methods, which will be discussed in detail in Part III.

Open-pollinated populations of outbreeding crops, such as corn, rye, the brassicas, and many forage species, have until very recently changed little in general genetic structure subsequent to domestication. The reason is that nearly all individuals in such populations have a different multilocus genotype; moreover, in each generation, nearly all individuals are likely to intercross with several other individuals, and the resulting ongoing extensive segregation and recombination breaks up favorable multilocus combinations of alleles. Segregation and recombination are also likely to insert large numbers of novel, untried genotypes into such populations each generation. Improvement of open-pollinated outcrossing populations depends on increasing the frequency of favorable alleles and favorable multilocus genotypes and, at the same time, maintaining past gains. Genetic uniformity in mass-selected outbreeding populations is difficult, and trueness-to-type for adaptedness and high performance are features of the population as a whole and rarely characteristics of individual plants. In the late decades of the nineteenth century and in the early decades of the twentieth century, corn breeders recognized that their populations were responding hardly at all to conventional mass-selection procedures (see Figure 5-1). This observation stimulated interest in alternative

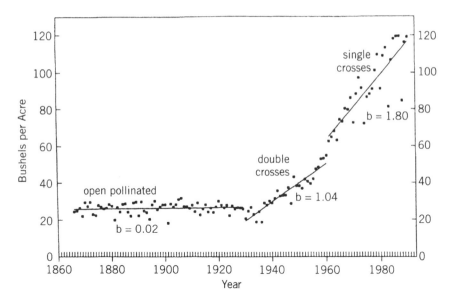

FIGURE 5-1    U.S. corn yields and kinds of corn—1866 to present; b values (regressions) indicate gain in bushels/acre/year (A.F. Troyer 1990).

methods of population improvement, particularly in the development of $F_1$ hybrid varieties. Such methods are discussed in detail in Chapter 14, but here we note only that the methods that have been adopted with outbreeding crops have often featured the development of monogenotypic $F_1$ hybrid varieties. More generally, the most successful population structure has featured a single genotype, a single pure line in the case of inbreeders, a single monogenotypic $F_1$ hybrid in the case of outcrossers, or either a single monogenotypic vegetatively propagated clone or a mixture of mostly superior vegetatively propagated clones (as is the case with sugar cane). Note also that hybrid varieties and clones can exploit all types of genetic variability, including both true single-locus overdominance (if it exists) and over-dominance, as well as epistasis. Homozygous inbred lines can exploit all types of genetic variability, except true overdominance (which appears to be rare). Mass-se-lected open-pollinated varieties can exploit all types of genetic variability, but not at maximum levels; heterogeneity mandates that some constituent genotypes in such populations must be below optimum in value. There is some evidence, however, that favorable interactions at the phenotypic level may sometimes com-pensate for the inferiority of some constituent genotypes in mixed populations.

# PART II

*Biological Foundations of Plant Breeding*

# SIX

# *Heredity and Environment*

Gregor Mendel (1865) made two major contributions to biology. First, he established the primary principle of genetics, namely, that heredity is particulate: it operates through highly constant hereditary particles, alleles that exist in two or more alternative forms for each hereditary element (gene). Only one member (allele) of each element (gene) is transmitted in each gamete. Different members of each element do not blend when they come in contact in zygotes, but they retain their integrity in passage from generation to generation. Alternative members of different elements follow simple algebraic rules in entering and emerging from offspring. The discovery of orderliness in inheritance—where none had been recognized previously—was of major importance in biology. Mendel identified three determining elements of inheritance and related them to the phenotypes they produce. In so doing, he established the critical distinction between genotype and phenotype. However, Mendel's second contribution, the experimental method by which he established his principles and by which the general applicability of his principles could be extended and tested, was perhaps equally important in the future development of biological sciences. Mendel's experiments had been deliberately designed to test a clearly stated theory. His experimental method was simple; it was novel in biology, and it provided a kind of experimental design that was remarkably efficacious.

Mendel's experimental method required that differences in any trait under investigation be large enough so that any individual could be assigned unambiguously to one or another of alternative classes. For example, Mendel's "short" peas ranged in height from 25 to 50 cm and his "tall" peas from 185 to 230 cm. Thus, despite the substantial variation within shorts and talls, there was no doubt whether any plant in parental, $F_1$, $F_2$, or other generations was either short or tall. Mendel analyzed this type of variation, which came to be called *discontinuous variation*, for seven traits in peas and showed that variation followed his rules of transmission in each and every case. Although gradations were observed within classes (e.g., within the short and tall classes), such within-class differences appeared to be continuous. Mendel was aware of continuous variation in his materials, but he ignored it, presumably because he recognized that this type of variation might complicate his analyses and thwart his main purpose, which was to determine the inheritance of discrete differences.

Continuous and discontinuous variation had been known in biology for many centuries before Mendel, and both types of variation had been observed in many different organisms. Darwin (1859 and later), for example, had painstakingly documented both continuous and discontinuous variation within wild populations as well as in domesticated plants and animals. A common observation in regard to continuous variation was that offspring tended to be intermediate between their parents and that further "blending" occurred in subsequent generations. This perception had led much earlier to the blending theory of inheritance, which postulated that the hereditary potential of parents blended in offspring and that the components of the blend did not regain their purity when transmitted to the next generation. However, it was known that some unusual variants, which were nonoverlapping (e.g., some color variants, brachytic dwarfs), sometimes appeared suddenly and that such discrete variants often persisted for many generations in some familial lines. This puzzling behavior was inconsistent with "blending inheritance."

In the eighteenth and nineteenth centuries it had been observed that selection for uniformity of type for both continuously varying and for more or less discretely differing characters was especially successful in cattle-breeding programs involving close matings (e.g., brother-sister, parent-offspring matings, and other types of close inbreeding). This result suggested that both continuously and discontinuously varying characters were partly heritable. During the nineteenth century pure-line selection had become well organized as a plant-breeding method, and it had become evident that many continuously varying traits had a heritable component.

Late in the nineteenth century, Francis Galton founded a group in England that sought to explain the inheritance of continuously varying differences, especially in humans but also in other species. Galton (1889) had found (in humans) that taller individuals had taller offspring, on the average, than shorter individuals and, hence, that such measurable (metrical characters) were partly heritable. To measure the degree to which such characters were inherited, new biometrical techniques (correlation and regression) were invented. The analyses of Galton and his followers indicated that offspring had mixtures of parental characteristics, and their offspring in turn had mixtures of characteristics derived from grandparents. When Mendel's work was brought to light in 1900, Galton's followers (the biometricians) were understandably reluctant to accept simple Mendelian ratios, and the discontinuous mode of inheritance they implied, as anything more than unusual and trivial exceptions to continuous or blending inheritance, which was seemingly much more common. The biometricians failed to grasp the essential distinction between the determinant (the genotype) and its effect (the phenotype). On the other hand, some of the early Mendelians also failed to grasp this distinction, arguing that continuous variation was never heritable. It remained for Wilhelm Johannsen (1903) of Denmark to provide a Mendelian explanation for the inheritance of continuously varying characters.

Johannsen accepted as credible the substantial evidence provided by pure-line plant breeders, by animal breeders who practiced line breeding, and by Galton and his followers, that continuous variation is partly heritable. Johannsen accomplished for continuously varying characters what Mendel had done 38 years earlier for discontinuous variation—he spelled out precisely the distinction between genotype and phenotype for such traits. He succeeded where others had failed. Johannsen, as Mendel before him had done for discrete traits, reduced the problem to its essence (many constant elements present in differing combinations), developed a clear and straightforward experimental design to test this proposition, and applied his design to well-chosen experimental materials. In designing his experiment, Johannsen took advantage of the fact that inbreeding increases homozygosity at all Mendelizing loci, regardless of the kind or magnitude of their effect on survival value. In particular, he took advantage of the fact that uniparental reproduction (selfing) rapidly leads to allelic fixation at all Mendelizing loci, as well as to rapid separation of all selfing populations into numerous genetically distinct homozygous families, each genetically uniform within itself. Johannsen chose as his experimental material a commercially obtained seed lot of the Princess variety of the common bean (*Phaseolus vulgaris*), a species that shares with Mendel's peas the property of nearly complete self-fertilization. He expected that virtually all seeds of his original seed lot would have been rendered homozygous at all or very nearly all of their "Mendelian" loci during the many generations in which the Princess variety had been reproduced by selfing in commercial production. All seeds and their progeny, in turn, should thus be genetically like their parents, and genetically like one another to the same extent. However, many or most of the progeny lines in the Princess variety were expected to differ from one another, because bean plants outcross occasionally and new combinations among Mendelian elements would be expected to segregate out of hybrids whose parents had come to differ over time in seed weight owing to spontaneous production of new genetic variability. Johannsen's design called for precise experimental measures of differences and similarities in seed weight between parents and offspring to determine the extent of variation that was heritable versus nonheritable. Johannsen's experiments led to the "pure-line theory," published in 1903, 37 years after Mendel's trailblazing paper but only 3 years after the rediscovery of Mendel's paper in 1900.

## THE PURE-LINE THEORY

Johannsen observed that there were seeds of many sizes in his original seed lot of the Princess variety. In his earliest experiment Johannsen (1903) weighed many seeds, grouped these seeds into 10-centigram weight classes, and grew progenies from each weight class. He found that progenies descended from the heavier parental seeds consistently had heavier seeds, on average, than progenies descended from lighter seeds. This result established that selection for this continuously varying trait had been effective, a result that was in conformity with the results of the biometricians and with Mendel's results.

Johannsen next established 19 lines by growing progenies descended from 19 different single seeds he had selected (Johannsen 1903) from his original seed lot. Detailed studies of these 19 lines revealed that each line had a characteristic mean seed weight. The mean seed weight of the heaviest line (Line 1) was 64.2 cg, and a regularly graded series of mean seed weights occurred from this value down to

35.1 cg, the mean seed weight of the lightest line (Line 19). No sharp between-line differences in mean seed weight were observed (the mean difference between the 19 lines was 1.53 cg). Johannsen concluded that the original commercial seed lot had consisted of a mixture of "pure lines," that each line was homozygous in the Mendelian sense at all or very nearly all Mendelian loci as a consequence of many generations of descent by uniparental reproduction during the numerous generations the Princess variety had been grown commercially. Measurements of seed weights within each of the 19 lines showed that variability in within-line seed weights was much smaller than variability in the original seed lot. Johannsen attributed the within-line variability primarily to differences in various environmental factors (e.g., the position occupied by individual seeds within the pod; seeds at the extreme ends of the pod were often smaller than other seeds). This interpretation could be readily checked experimentally. As a first check, seeds within each of the 19 lines were separated into 10 cg weight classes and progenies were grown from each weight class. Johannsen found that all classes within each line produced progeny of the same mean seed weight; for example, the four categories centered on 20, 30, 40, and 50 cg that occurred in Line 13 (mean seed weight 45.5 cg) produced progenies with statistically nonsignificant differences in mean seed weight of 47.5, 45.0, 45.1, and 45.8 cg, respectively. Within each line, seeds of the various weights all produced progenies with mean weights characteristic of the line, within sampling errors. Further evidence was obtained by selecting the heaviest and lightest seeds within each line for six successive generations. Such selection was convincingly ineffective; for example, after six generations, in Line 1 the mean weights of the light and heavy selection lines were 69 and 68 cg, respectively. The mean weight of each line remained remarkably constant generation after generation, indicating that very little of the variability within lines was heritable.

Johannsen also obtained an estimate of the proportion of variability in seed weight resulting from heredity, and the proportion resulting from nonheritable causes, by calculating parent-offspring correlation coefficients within lines and also within a mixed population made up of all 19 lines. If variation is at least partly under genetic control, it would be expected that heavy-seeded parents would produce heavy-seeded offspring, and light-seeded parents light-seeded offspring; in other words, parent-offspring correlations are expected to be positive and statistically significant. If little or none of the variability is due to genetic causes, parent-offspring correlations would be expected to be statistically nonsignificant. Johannsen reported a positive and statistically highly significant parent-offspring correlation of $r = 0.336 \pm 0.08$ for the mixed population, indicating that about one-third of the total variation in seed weight in the mixed population was due to genetic differences between lines and about two-thirds resulted from nongenetic causes. However, parent-offspring correlations were nonsignificant within the lines; for example, $r$ was $0.018 \pm 0.038$ within Line 13, indicating that effectively all of the variability was due to nongenetic causes, as expected in highly homozygous "pure lines." Seed weight in mixed populations, such as the Princess variety, showed heritable and nonheritable variation simultaneously (as has subsequently been found to be the case for most measurement characters in most populations). Clearly, however, a properly designed breeding test such as Johannsen's was needed to distinguish between the two types of variability. Parent-offspring regressions, often estimated visually, have subsequently proved to be a highly useful tool for plant breeders.

## THE SIGNIFICANCE OF JOHANNSEN'S EXPERIMENTS

Johannsen's experiments provided the basic facts needed to clarify the attributes of selection for continuously varying traits and the consequences of such selection. Common beans are nearly entirely self-fertilizing, and because Johannsen's original seed lot had been propagated for many generations in commercial plantings, nearly all seeds were expected to be homozygous at nearly all genetic loci. Selection within progenies derived from single seeds was consistently unsuccessful in changing seed weight. This established that the same primary hereditary elements were not only present in each generation but present in the same combinations generation after generation; hence, all individuals within each line, in common with Mendel's traits, had the same potential for development. This result indicated that the genotype of each line had remained constant throughout Johannsen's experiments. Yet, at the same time, mean seed weight differed among lines derived from different seeds of the original seed lot, indicating changeableness. There were, however, no conspicuous discontinuities between lines, indicating that few, if any, of the genetic elements governing seed weight had major effects on seed weight, even though the cumulative effects of the elements of each line were large (64.2 − 35.1 = 29.1 cg) in the particular 19 lines Johannsen had studied in detail. Hence, Mendelian ratios could not have been detected even if elements governing seed weight had segregated in Mendelian fashion. Johannsen concluded that the properties of the elements governing continuously varying traits were parallel to those of the Mendelian elements that govern discontinuous variability. Mendel had shown that hereditary elements governing discontinuous variation were similarly constant, emerging unchanged (unblended) in the progeny of hybrids descended from parents carrying different elements, while at the same time differences in weight resulted when the elements were present in different combinations in the progeny. Johannsen reasoned that constant elements appearing in differing combinations were also responsible for the apparently antithetical constancy versus variability that he had found for seed weight in beans. Thus Johannsen's experiments established that Mendelian inheritance is not limited to cases in which it can be detected by the methods used by Mendel; it may, in fact, be responsible for all heritable variation. A wide spectrum in the effects of single elements could now be imagined. At one extreme, some elements (such as those used by Mendel to deduce his principles) had large enough effects that they stand out from other elements and from nonheritable agents. At the other extreme, there were also elements with effects so slight that the differences they caused were always blurred by the effects of other agents, both heritable and nonheritable. The effect of a single minor element might entirely escape detection, but several to many such elements reinforcing one another might produce large differences in total. Johannsen's results made it clear to nearly all investigators that the ability to detect a heritable element by Mendel's methods did not depend on its mode of inheritance but, instead, on the kind and magnitude of its effect on appearance and/or measurability.

Johannsen's contributions in regard to continuously varying traits were thus parallel to those of Mendel regarding discretely inherited traits. Mendel gave us an experimental method for analyzing discontinuous variation and demonstrated its power by showing how such variation can be ascribed unambiguously to Mendelizing hereditary units. Johannsen gave us a different experimental method for analyzing continuous variation and demonstrated its power by showing how such variation, although not susceptible to analysis by Mendel's methods, could nevertheless be attributed to multiple Mendelizing units, each with a small individual effect. Johannsen also gave us one of the most fundamental concepts in all of biology, and certainly one the most useful concepts in plant breeding—the relationship between genotype and phenotype for continuously varying traits. Our ideas about, and terminology for, continuously varying traits are attributable to Johannsen (1903, 1913). The genotype (total genetic composition) determines the potential for development for each individual, whereas the environment in which the individual lives determines the outcome of development (the observable and/or measurable feature or features of the phenotype). Many, indeed most, of the issues that arise in plant breeding can be traced back to the distinction between genotype and phenotype. The traits that are most important in plant breeding (such as productivity, product quality, size, growth rate, and stress resistance) are nearly always continuously varying traits. The objective of breeding is to construct superior genotypes. But how can breeders, working with genes that usually cannot be followed individually, develop genotypes whose superiority cannot be recognized by mere inspection of its phenotype? Much of the rest of this book will be concerned with attempting to answer this question.

## THE MULTIPLE-FACTOR HYPOTHESIS

Johannsen's experiments (1903, 1913) established that the Princess variety of beans consisted of a mixture of pure lines and that, although variations in seed weight within pure lines were nearly entirely nonheritable, differences in seed weight between individual pure lines were highly heritable. This led Johannsen to postulate correctly that between-line differences in seed weight were due to Mendelizing units. However, variations in seed weight within pure lines were small, causing overlapping of the seed weight ranges of different genotypes. By definition, therefore, genes that do not produce phenotypic discontinuities cannot be analyzed by Mendel's methods. Consequently, Johannsen was unable to identify any specific Mendelian units, and, largely for this reason, his work failed to resolve the dispute concerning the nature of continuous variation. However, Johannsen's results had established that continuous variation was partly heritable and that if single gene differences were small, as compared with nonheritable variation, such genes could not be followed individually using Mendel's methods. What was needed was a method of handling the genes for continuous variation en masse. G.V. Yule, an English biometrician, and H. Nilsson-Ehle, a Swedish plant breeder, provided the ideas that led to a successful methodology. Yule (1906) suggested that there need be no conflict between Mendel's particulate inheritance and the inheritance of continuously varying traits, provided many genes having similar small effects were responsible for continuously varying traits. About the same time, Nilsson-Ehle reported a naturally occurring model for the type of inheritance suggested by Yule. In bread wheat (*Triticum aestivum*) and cultivated hexaploid oats (*Avena sativa*) it was known that three Mendelian loci governed red versus white kernel color. Any one of the three loci gave a ratio of 3 red to 1 white when segregating alone. Two loci segregating, together gave a 15:1 ratio, and all three loci a ratio of 63:1. $F_1$ hybrids were intermediate between their parents in intensity of redness, and in the $F_2$ generation various intensities of color were observed. Nilsson-Ehle (1909), using progeny tests, was able to associate the color differences with differences in the number of alleles for redness. In crosses between white and deep-red parents, the $F_1$, which carries 3 recessive white and 3 dominant red alleles, was intermediate between the parents, and the $F_2$ distribution indicated a ratio of 1:6:15:20:15:6:1 for $F_2$ individuals carrying 0, 1, 2, 3, 4, 5, or 6 red alleles. Thus, with as few as three loci segregating, the distribution approached a normal curve. This established that the effects of different alleles of different loci could be cumulative. Thus, as suggested by Johannsen and Yule, Mendelian genes can have properties necessary to account for the inheritance of continuous variation. Much more compelling evidence that plural segregating genes with similar and cumulative effects on the phenotype can explain the inheritance of quantitative traits was provided in 1916 by E. M. East (United States). We now examine one of East's experiments in some detail because it reveals so convincingly additional features of the type of inheritance that is of primary importance in plant breeding.

## COROLLA LENGTH IN *NICOTIANA LONGIFLORA*

Corolla length in *Nicotiana longiflora* is a particularly favorable trait for study of the inheritance of continuous variation, because the corolla length of any variety is nearly constant under a wide variety of environmental conditions. For this reason East hybridized two true-breeding, varieties of this species, one with extreme short and the other with extreme long corollas, for one of his investigations of the nature of quantitative inheritance, obtaining the data summarized in Table 6-1. It was immediately apparent from these data that variability in corolla length was continuous, thus precluding a standard Mendelian analysis. East postulated, however, that although part of the variation in corolla length was nonheritable, most was under the control of Mendelian genes. East's reasoning was as follows: The parents differed strikingly in corolla length when they were grown under similar conditions. The difference was real, and its basis lay in Mendelian differences; the parents had been long, inbred and must have approached complete homozygosity. However, there were small plant-to-plant differences in corolla length in the parents, and this variability presumably resulted from subtle environmental differences within the experimental garden. Corolla length of the $F_1$ was intermediate between the parents, and variability within the $F_1$ was comparable to that within the parents. These observations conform to Mendelian expectations: intermediacy if the Mendelian genes act additively, and $F_1$ variability similar to that of the parents because all $F_1$ plants in a hybrid between two pure lines are expected to be identical genetically. The frequency distributions, standard errors (s.e.), and coefficients of variability

TABLE 6-1 Frequency distributions for corolla lengths in parents, $F_1$, $F_2$, and succeeding generations in *Nicotiana longiflora* (East 1916)

Class Centers in Millimeters

| Generation | Parent Size | 34 | 37 | 40 | 43 | 46 | 49 | 52 | 55 | 58 | 61 | 64 | 67 | 70 | 73 | 76 | 79 | 82 | 85 | 88 | 91 | 94 | 97 | 100 | n[1] | $\overline{X}$[2] | SE[3] | CV[4] |
|---|---|---|---|---|---|---|---|---|---|---|---|---|---|---|---|---|---|---|---|---|---|---|---|---|---|---|---|---|
| $P_1$ |  | 1 | 21 | 140 | 49 |  |  |  |  |  |  |  |  |  |  |  |  |  |  |  |  |  |  |  | 211 | 40.4 | 1.6 | 4.0 |
| $F_1$ |  |  |  |  |  |  |  |  |  | 4 | 10 | 41 | 40 | 3 |  |  |  |  |  |  |  |  |  |  | 173 | 63.5 | 2.9 | 4.6 |
| $P_2$ |  |  |  |  |  |  |  |  |  |  |  |  |  |  |  |  |  |  |  |  | 13 | 45 | 91 | 19 | 159 | 93.1 | 2.6 | 2.8 |
| $F_2$ |  |  |  |  |  |  | 3 | 9 | 18 | 47 | 55 | 93 | 75 | 43 | 25 | 7 | 8 | 1 |  |  |  |  |  |  | 444 | 68.7 | 6.4 | 9.2 |
| $F_3$ | 46 |  |  |  | 1 | 4 | 26 | 44 | 38 | 22 | 7 | 1 |  |  |  |  |  |  |  |  |  |  |  |  | 143 | 53.5 | 3.7 | 7.0 |
| $F_3$ | 50 |  |  |  | 6 | 20 | 53 | 49 | 15 | 4 |  |  |  |  |  |  |  |  |  |  |  |  |  |  | 147 | 50.2 | 3.2 | 6.3 |
| $F_3$ | 60 |  |  |  | 2 | 3 | 9 | 25 | 37 | 70 | 19 | 10 |  |  |  |  |  |  |  |  |  |  |  |  | 175 | 56.3 | 4.1 | 7.2 |
| $F_3$ | 72 |  |  |  |  |  |  |  |  |  | 4 | 20 | 25 | 59 | 41 | 19 | 2 |  |  |  |  |  |  |  | 170 | 70.1 | 3.8 | 5.2 |
| $F_3$ | 77 |  |  |  |  |  | 1 |  | 1 | 1 | 1 | 2 | 16 | 33 | 43 | 34 | 20 | 6 | 1 |  |  |  |  |  | 159 | 73.0 | 5.0 | 6.9 |
| $F_3$ | 80 |  |  |  |  |  |  |  |  |  | 2 | 8 | 14 | 21 | 39 | 39 | 32 | 10 | 1 |  |  |  |  |  | 166 | 74.0 | 4.9 | 6.6 |
| $F_3$ | 81 |  |  |  |  |  |  |  |  |  | 1 | 1 | 8 | 16 | 20 | 32 | 41 | 17 | 3 | 3 | 1 |  |  |  | 143 | 76.3 | 5.1 | 6.6 |
| $F_3$ | 82 |  |  |  |  |  |  |  |  |  |  |  | 3 | 5 | 12 | 20 | 40 | 41 | 30 | 9 | 2 |  |  |  | 162 | 80.2 | 4.8 | 5.9 |
| $F_4$ | 44 |  |  | 8 | 42 | 95 | 38 | 1 |  |  |  |  |  |  |  |  |  |  |  |  |  |  |  |  | 184 | 45.7 | 2.4 | 5.2 |
| $F_4$ | 43 |  |  | 2 | 23 | 122 | 41 | 1 |  |  |  |  |  |  |  |  |  |  |  |  |  |  |  |  | 189 | 46.3 | 1.9 | 4.0 |
| $F_4$ | 85 |  |  |  |  |  |  |  |  |  |  |  |  |  | 4 | 9 | 38 | 75 | 59 | 6 | 3 | 1 |  |  | 195 | 82.2 | 3.3 | 4.0 |
| $F_4$ | 87 |  |  |  |  |  |  |  |  |  |  |  | 14 | 5 | 6 | 11 | 21 | 23 | 41 | 29 | 8 | 5 | 1 |  | 164 | 82.9 | 5.8 | 7.0 |
| $F_5$ | 41 | 3 | 6 | 48 | 90 | 14 |  |  |  |  |  |  |  |  |  |  |  |  |  |  |  |  |  |  | 161 | 42.0 | 2.3 | 5.5 |
| $F_5$ | 90 |  |  |  |  |  |  |  |  |  |  |  |  |  | 2 | 3 | 8 | 14 | 20 | 25 | 25 | 20 | 8 |  | 125 | 87.9 | 5.2 | 6.3 |

[1]No. plants scored. [2]Mean corolla length. [3]Standard error. [4]Coefficient of variability = S.E./X.

(c.v.) for $P_1$, $P_2$, $F_1$, and $F_2$ left no doubt that the $F_2$ generation was much more variable than the parental and $F_1$ generations. The $F_2$ generation could not be expected to be immune from environmental effects, but neither could it be expected to be dramatically more sensitive than other generations. Consequently, the much greater variability of the $F_2$ cannot reasonably be accounted for on the basis of environmental effects alone. Segregation and recombination among plural Mendelian genes, producing many new allelic combinations, is expected in the $F_2$ generation. East attributed the great increase in variability in that generation to Mendelian segregation and recombination.

The patterns observed in the parental, $F_1$, and $F_2$ generations appeared to conform to Mendelian expectations, based on several segregating genes; this raised the question of the precise number and nature of the genes involved. If all of the variability owing to Mendelism were controlled by a single allelic pair, three genotypes ($a_1a_1$, $a_1a_2$, $a_2a_2$) would be expected in the $F_2$ generation and three phenotypes corresponding to the phenotypes of the parents and $F_1$ (Table 6-1) should be associated with these genotypes; that is, about 1/4, 1/2, 1/4, respectively, should fall into the range of about 34–43, 58–70, and 91–100 mm. Thus all of the 444 $F_2$ plants should fall into one of three discrete groups. Table 6-1 shows that a single-locus hypothesis cannot account for the approximately normal $F_2$ distribution curve actually observed. Suppose, however, that the genetic differences were due to two allelic pairs instead of one. With no dominance at either locus, if each of the four alleles have equal and cumulative effect, the $F_1$ plants should have corollas approximately 67 mm long, on average, and there should be five classes in the $F_2$ plants with means of approximately 40, 54, 67, 81, and 93 mm. The five classes should appear in a ratio of 1/16:4/16:6/16:4/16:1/16. The means of these five phenotypic classes should be separated by approximately 13.5 mm and, because the parents and the $F_1$ varied over ranges of approximately 9 to 15 mm, the classes should be discrete or very nearly so. It should be possible to place nearly all individuals into their proper genotypic class with only slight possibility of misclassification. Although a two-locus distribution constructed on this basis clearly does not fit the nearly normal curve actually observed, the assumption of an extra allelic pair seems to be a step in the right direction. Constructing curves based on three, four, or more allelic pairs, assuming more or less equal and cumulative effects of each locus, shows that expected frequencies approximate more and more closely the normal frequency distribution curve actually observed.

Examination of the $F_2$ frequency distribution shows that only one $F_2$ individual came within a single size class of the larger parent and that no $F_2$ individuals came within two size classes of the upper size limit of the smaller parent. Thus, in an $F_2$ population of 444 plants, the parental genotypes would not be expected to be recovered. With four loci involved, and assuming that previous assumptions apply, there should be about an even chance of recovering the parental types in an $F_2$ of the size studied. Consequently, it is reasonable to assume that more than four loci were involved. If only five loci were involved, and previous assumptions apply, the means of the expected classes are expected to differ by less than 5 mm. However, environmental effects are expected to cause any single genotype to vary over a range of about 9 to about 15 mm. It is thus readily apparent that genotypic classes would overlap phenotypically and, hence, that the distribution curve would approach smoothness. Although no hypothesis is possible regarding the precise number of genes and/or alleles involved, or their exact mode of action, the observed results are of the type expected if several segregating genes, possibly as few as five, govern corolla length in this particular $F_1$ hybrid.

This scheme fits the observed facts neatly, but it does not establish beyond doubt that plural segregating genes are the correct explanation for the variation observed. Supporting evidence would be reassuring, and this East obtained through predicting the consequences of such an explanation in the $F_3$ and subsequent generations by testing various predictions against observations. East's first additional prediction (Prediction 5) was that individuals from different parts of the $F_2$ distribution curve should, in later generations, produce progenies differing markedly in mean corolla length; that is, that parent-offspring correlations should be positive, as had been the case with Johannsen's beans. Data from the $F_3$, $F_4$, and $F_5$ generations, also summarized in Table 6-1, made it possible to test this prediction. Clearly, many $F_3$ families differed significantly in corolla length, and the differences were positively correlated with the corolla length of the $F_2$ parent in each family. $F_2$ plants with short corollas produced $F_3$ families with shorter corollas than $F_2$ plants with longer corollas. The same was the case in the transitions from generation $F_3$ to $F_4$ and from generation $F_4$ to $F_5$.

Prediction 6 was that different $F_2$ individuals might, owing to sampling accidents during selfing, produce progenies differing in heterozygosity and, hence, in variability. The possible range of variabilities should extend from progenies as variable as the $F_2$ (expected to be rare even in generation $F_3$) to progenies no more variable than the original parents (expected to appear only after three, four, or more generations of selfing). In generations succeeding the $F_2$, the variability of a family could be expected to be the same or less than the family from which it came, but not greater (Prediction 7). Data from generations $F_2$ through $F_5$ allowed tests of these predictions. It can be seen, both from the frequency distributions and from their coefficients of variability, that $F_3$ and later-generation families differed widely in variability. No $F_3$ families comparable in variability to either parent appeared— the number of $F_3$ families grown was rather too small to expect this to have occurred. However, two of the four $F_4$ families were comparable to the short parent.

Because homozygosity within families should increase with self-fertilization, genetic variability within each succeeding generation is expected to decrease. It is apparent that the coefficient of variability in the $F_2$ generation was much higher than that in any of the $F_3$ families and that this decreasing trend continued into the $F_4$ generation. In a side experiment, East also practiced selection for the longest and shortest corollas in each generation from $F_2$ through $F_5$. He obtained one $F_5$ family that approached the shorter parent in corolla length and one that approached the longer parent. This provided supporting evidence that heterozygosity for genes affecting corolla length persisted into the $F_5$ generation. East had selected as parents the shortest and longest floral types known in *N. longiflora* and so did not expect transgressive corolla-length variants in segregating generations.

It was from experiments such as those of Johannsen, Nilsson-Ehle, and East that the Mendelian scheme was established as capable of explaining the inheritance of continuously varying characters. This explanation for the inheritance of quantitative characters came to be called the *multiple-factor hypothesis*. Although the multiple-factor hypothesis viewed the genotypes for a given quantitative trait as simply the sum of the effects of the several genes postulated to be involved, and did not visualize the interacting and coordinated genetic systems now regarded as basic to quantitative inheritance, it was one of the most important contributions to genetic thought.

## ORIGIN OF GENETIC VARIATION

"Pure lines" have thus far been discussed as fixed and unalterable entities, and there is a great deal of experimental evidence that this viewpoint is a close approximation to reality in small, short-term experiments such as those of Johannsen and East. At the same time, long before Johannsen, there had been many reports of inherited variations that had arisen de novo in populations of many different species of plants and animals. The evidence was especially clear in established, true-breeding varieties of barley, wheat, peas, beans, and other heavily self-pollinated plants in which sudden heritable variants affecting many different traits, both discrete and qualitative (e.g., flower and/or seed color), as well as continuous or quantitative traits (time of flowering, tillering capacity), had been reported. Further, the great diversity of types documented by Darwin and others as occurring in all species and often fixed in different races or breeds, provided evidence for the long-term importance of spontaneously arising heritable variants in producing genetic variability. We now know that these variants, both qualitative and quantitative, have their origin in mutation and that their appearance in numerous combinations with other mutants has come about through hybridization and its Mendelian consequences within populations, as well as from migrations among populations. An appreciation of these processes is important in plant breeding, because naturally occurring variation has been the main source, indeed almost the exclusive source, of the natural heritable variability on which improvement of cultivated plants depends. We now consider these processes to set the stage for more detailed discussions in later chapters.

East was perhaps the earliest researcher to obtain estimates of mutation rates in plants. His evidence came from matriclinous individuals that arose during attempts to produce species hybrids in *Nicotiana*. These individuals, which presumably arose parthenogenetically from unfertilized eggs, were diploid and presumably completely homozygous for the genotype of the maternal parent. According to East, the progenies of these plants were remarkably alike. However, East commented that after as few as four generations of self-fertilization, the progenies of these plants were nearly as variable as ordinary inbreds. Hybridizations and mixtures had been excluded beyond reasonable doubt, which led East to the conclusion that the increase in variability was due to mutations and, hence, that spontaneous mutation rates must be quite high. It remained for Stadler (1942) to obtain the first precise and convincing evidence of the frequencies of specific gene changes in plants. Stadler selected favorable materials for study, namely, several well-known loci governing endosperm traits in corn—for example, $R$ (color factor), $I$ (color inhibitor), $P_2$ (purple) $Su$ (sugary), $Y$ (yellow), $Sh$ (shrunken), and $Wx$ (waxy). Stadler scored large numbers of endosperms (~265,000 to ~2,500,000) and observed very different mutation rates for different loci (e.g., 492 mutations per million gametes for $R$, 106 mutations per million gametes for $I$, 2.4 mutations per million gametes for $Su$, 2.2 mutations per million gametes for $Y$, 1.2 mutations per million gametes for $Sh$, and none in 1,503,744 gametes tested for $Wx$). Stadler concluded that mutation rates vary widely from locus to locus. More recent studies suggest that usual mutation rates in higher plants vary from about $1 \times 10^{-5}$ to $1 \times 10^{-7}$/locus/generation, although lower and higher rates have been reported. In barley, Kahler et al. (1984) observed no mutants in 3,386,850 allele replications for four well-known loci governing morphological traits and color ($Vv$, two-rowed vs. six-rowed spikes; $Bbbl$, blue vs. white color; $Rr$, rough vs. smooth awns; $Oo$, white vs. orange lemma color). This result indicates that the upper bound for the mutation rate at these loci is $8.85 \times 10^{-7}$ per generation. These authors also estimated spontaneous mutation rates by assaying 84,126 seedlings of a highly homozygous barley line for five enzyme loci (esterase loci 1, 2, 3, and 4; and acid phosphatase locus 1). No mutants were observed in 821,260 allele replications, a result that excludes ($p = 0.95$) spontaneous mutation rates higher than $3.56 \times 10^{-5}$ for these enzyme loci.

Theory indicates that chances are few that any new mutant (including adaptively beneficial mutants) will become established in the population in which it occurs (Fisher 1930). The probability that dominant mutants that have serious deleterious effects on survival will become established is, for obvious reasons, vanishingly small. This is also the case for seriously debilitating recessive mutants, especially in heavily inbreeding populations (Wright 1937), because inbreeding quickly leads to homozygosity, exposing such recessives to rapid elimination by selection. In outbreeding populations such recessive mutations have the advantage of "hiding" in heterozygotes and, thus, may escape elimination longer. However, adaptively neutral and even adaptively beneficial alleles are also likely to be lost, especially in the first few generations after their initial appearance, when they are rare in the population and particularly subject to loss owing to chance events (genetic drift). Despite the low probability that any single mutant will survive, the great diversity of types present in nearly all populations provides strong evidence concerning the long-term importance of mutations in creating variability. The large numbers of novel types, not present in wild ancestors, that have appeared in cultivated plants show that genetic variation arises de novo at rates that can assume importance in plant breeding. This is not surprising, for three reasons. First, populations of plants in cultivation are often very large, many millions of individuals per generation, so that even though spontaneous mutation rates are low, substantial numbers of mutants may occur per locus/year. Second, agricultural environments differ in many respects from the environments occupied by wild ancestors, providing opportunities for novel alleles better adapted to cultivated environments to replace ancestral alleles that had evolved in the wild. Third, prehistoric cultivators, as well as their modern successors, were careful observers who almost certainly saved many agriculturally desirable variants that would probably not have survived in the wild.

Although there is a growing body of evidence suggesting that many "spontaneous" mutations result from insertion of transposons into, or close to, particular genes, it is adequate for present purposes to consider "point" mutations to be spontaneously inherited chances in the base sequences of the DNA of an organism.

Not all mutations lead to phenotypic changes that can be detected by presently available techniques—for example, new alleles affecting quantitative traits are difficult to detect. However, most mutations that cause phenotypically detectable changes make the individuals that inherit such mutations less fit than individuals with the original base sequence and phenotype. It is not surprising that mutations almost always have deleterious effects—extant base sequences and the enzymes they encode are the products of long continued evolutionary processes, and it seems unlikely that random accidental chances in amino acid sequences would often improve the ability of such time-tested enzymes to carry out their functions. Because new mutants are nearly always deleterious, high mutation rates are likely to endanger the survival of highly mutable species. It has often been surmised that evolution adjusts the mutation rate of each gene to an appropriately low level, presumably because if there were too few new mutants in total, there would also be too few favorable mutants and the species would ultimately be unable to meet new challenges imposed by changes in its biological and/or physical environment.

# SEVEN

# *Genetic Consequences of Hybridization*

In previous chapters it has been seen that the natural hybridizations that occur within populations of cultivated plants, and their Mendelian consequences compounded over many generations, have played an important role in producing novel genotypes that have contributed to plant-breeding progress. However, both the range and diversity of recombinants that can be expected from natural hybridizations are obviously limited by the similarity of the types likely to hybridize, especially within populations of largely self-pollinated crops after hundreds of generations of selection in cultivation. It is not surprising, then, that returns from mass selection and pure-line breeding have become progressively smaller as the stores of genetic variability that have accumulated over the centuries in cultivated selfing plants have been exploited. As this happened, the emphasis in plant breeding shifted, especially during the first half of the twentieth century, to planned hybridizations between types with phenotypic characteristics that breeders believed would be most likely to promote progress, and by the mid-twentieth century, breeding methods increasingly featured artificial hybridizations between carefully selected parents.

In making hybrids between genotypes that possess between them characteristics that are likely to produce superior recombinants, breeders created populations in which selection was often very rewarding. Yet it also became apparent that some hybridizations led to disappointing failures. Fortunately, information has steadily increased and modern plant breeders are now in a much more favorable position than their predecessors to choose parents and to guide changes in allelic frequencies in desired directions. It is to this background of evolutionary genetic understanding that attention will now be directed.

## THE GENE-CHARACTER RELATIONSHIP

Alleles of many different loci are the building blocks from which new varieties are assembled, and it is essential in plant breeding to keep in mind the relationships between these many alleles and phenotypic traits, that is, the relationships between genotypes and phenotypes. It is now commonly accepted that few phenotypic effects are related to single alleles in any simple way. Rather, better phenotypes result from biochemical reactions initiated by specific alleles of many loci, through complex chains of events that are often molded and shaped by alleles of other genes and nearly always much affected by the external environment. In some cases specific morphological or physiological attributes appear to be controlled largely by single genes and to be little affected by either the genetic environment or the external physical environment. That is, important commercial characteristics of many crop species (e.g., determinant vs. indeterminate habit of growth, resistance to specific pathotypes of diseases) sometimes appear to be governed by alleles of single loci with large, easily recognizable, and sometimes stable effects. Such alleles are particularly valuable to plant breeders because they are readily managed by simple and straightforward Mendelian techniques. But even these familiar "qualitative" loci of classical genetics have frequently been found to have phenotypic manifestations seemingly unrelated to the primary effect of the locus. For instance, in replicated trials in which otherwise isogenic awned and awnless types of barley were compared, the awned isogenics often yielded significantly more than the awnless isogenics. Furthermore, comparable decreases (or increases) in yield sometimes attended removal of awns from an awned variety by backcrossing. Differences between otherwise isogenic pairs sometimes appear to depend on the marker alleles, sometimes on the background genotype, sometimes on the environment, and sometimes on all such factors. Isogenic lines can also be obtained by hybridizing individuals differing at one or more easily identified loci, selfing such hybrids, and selecting heterozygous individuals at each such identifiable locus in several successive generations. This enforces heterozygosity at the loci held heterozygous while inbreeding occurs at all other loci. After many generations each family can be expected to be homozygous and isogenic at all loci, except the one or more loci that are maintained in heterozygous condition. At this point the two homozygotes in question, in any family, can be extracted and any differences between them can reasonably be ascribed to the locus in question, or to very tightly linked loci. Such studies have shown that specific alleles of many different loci often have significant side effects. In some cases, however, the side effects in a given family have been different when the family was tested in two or more sets of environmental conditions; that is, genotype-environmental interactions also played significant roles. In other cases, the side effects of a given locus differed from one isogenic family to another; that is, interactions of given alleles with the differing background genotypes of different family lines can also play a significant role. Clearly, extensive testing may be required to determine whether apparent side effects of a "marker" allele are constant over environments and genetic backgrounds. In some cases side effects have been favorable, constant, and inseparable from a marker allele. Inability to disassociate side effects from the marker alleles after very large numbers of opportunities for recombination implies that such effects may result from pleiotropy rather than from linkage. In plant-breeding practice it makes little difference whether a consistently favorable side effect results from pleiotropy or linkage, although the genetic basis of the difference clearly is not the same in the two cases. When side effects are unfavorable, however, pleiotropy obviously prevents the utilization of major genes as an aid in plant breeding. This point can be illustrated with attempts to take advantage of the stem rust and mildew resistance of C.I. 12633, a bread wheat ($n = 21$ chromosomes) derived from *Triticum timopheevi* ($n =$

14 chromosomes) by repeated backcrossing to a widely grown bread wheat variety. The rust and mildew resistance of C.I. 12633 is governed by two loci (14.8% crossing over) located on chromosome XIII. No difficulties were encountered in transferring the resistance to both diseases into other wheat varieties, but resistant progeny consistently yielded about 5% less than susceptible progeny. When more than 20 additional opportunities for recombination in backcrosses (as well as in recombinant inbred lines) failed to break the association between resistance and inferior yielding ability, it was concluded that the association was either due to pleiotropy of the resistance alleles or, alternatively, that the chromosome segment (~10 crossover units long) in which the resistance genes were located carried one or more nonresistance loci with deleterious effects on yield. This result also suggested the possibility that crossing-over within this short alien chromosome segment that had been transferred from *T. timopheevi* to *T. aestivum* occurs only rarely or not at all.

## EPISTASIS

*Epistasis* is a term originally used by Bateson in 1909 to describe forms of interlocus interactions by which an allele at one locus interferes (stands over) the phenotypic expression of an allele (or more than one allele) of another locus so that the phenotype is determined effectively by the former and not by the latter when both types of alleles occur together in the genotype. Alleles with expressions that are altered by alleles of other loci are designated hypostatic (standing under). The terms *epistasis* and *hypostasis* have gradually acquired a more general meaning that is synonymous with interactions between alleles of different loci; that is, epistasis and hypostasis are now used to describe phenomena whereby the effects of alleles of one locus may change in various ways with the presence or absence of alleles at other loci. Epistasis, or interallelic interaction, should be strictly distinguished from dominance, which refers to interactions among alleles of the same locus (intralocus allelic interactions). Epistasis, which usually does not depend on heterozygosity, makes possible stable multilocus combinations of alleles that often make important contributions to adaptedness. Recent evidence, to be considered in more detail in later chapters, indicates that epistasis plays a much more important role in adaptedness in inbreeders than overdominance, and that it also plays a much more important role than overdominance in outbreeders.

The simplest epistatic situations are those leading to familiar dihybrid ratios such as 9:7, 13:3, 15:1, 9:3:4, and 12:3:1. When alleles of three loci interact, the possibilities are much greater, and ratios such as 37:27, 55:9, and 27:9:9:19 have been observed. Biochemical analyses indicate that a common pattern is one in which one locus produces a substrate that serves as a raw material for a second locus. HCN (hydrocyanic acid) content in white clover, first worked out in 1943, follows this pattern: one locus produces a cyanogenic glucoside and another the enzyme that catalyzes the release of HCN from the glucoside. Plants with at least one dominant allele at each locus are high in HCN because both the substrate and the enzyme are present. Individuals homozygous recessive at either or both loci are low in cyanide for lack of either the substrate, the enzyme, or both. Many recent studies of multilocus enzyme polymorphisms suggest that complex epistatic systems, often involving "supergenes" made up of three, four, or five, up to many loci, usually loci located on different chromosomes, can play a major role in promoting adaptedness. Such supergenes often play a critical role in evolution as well as in plant breeding; hence, they will be considered in more detail in later chapters, especially in Chapter 10.

## MODIFYING FACTORS

Another way in which the genetic environment can adjust the gene-character relationship is through alleles of different loci with generally small effects that exert their influence by intensifying or diminishing the expression of major genes. Such alleles are appropriately called *modifying factors*. In some instances it has been difficult to demonstrate that modifiers have any effect other than in changing the expression of some specific major gene. One of the more thoroughly studied cases is the mottling of the seed coats of certain annual legumes. First, whether a seed is mottled at all depends on a dominant allele of a major gene. However, the degree to which the seed is mottled often depends on several to many alleles of one or more other genes, each with a different effect on mottling. Variation can be continuous, ranging from no sign of mottling to intense mottling over the entire seed coat. For example, seeds of plants that are heavily mottled sometimes survive well, whereas selection for less intense mottling in cultivated types often leads to lower survival ability. Many major genes appear to have extensive complements of modifiers, which has led some breeders to predict that modifiers will become increasingly important as more and more major genes become concentrated in cultivated varieties; hence, these breeders have suggested that plant breeding may ultimately reach the point where gains will increasingly often depend on the manipulation of modifiers.

## PENETRANCE AND EXPRESSIVITY

Penetrance has to do with whether an allele is capable of expressing itself in all individuals that carry the allele. An old and widely grown variety (Ventura) of the lima bean *(Phaseolus lunatus)* is homozygous for a dominant allele that causes the tips and margins of unifoliate leaves of seedling plants to be partly deficient in chlorophyll. Yet it is rare for more than 10% of seedlings to show this characteristic; hence, the penetrance of this allele is usually less than 10%. Under certain environmental conditions, however, nearly all plants show the character so that penetrance is nearly 100%. In other environmental conditions no plants show the character, so that penetrance is zero. In addition, there can be wide variation in the manner in which these alleles are expressed in different plants. In some individuals unifoliate leaves are devoid of chlorophyll. The unifoliate leaves of some affected plants soon become yellow but shortly thereafter become normal green in coloration, whereas in other individuals the unifoliate leaves fail to develop chlorophyll and abscise. In still other individuals only the tip of the blade is affected, and in some only the margin of the leaf blade is affected. Whether an allele is expressed at all is denoted by the term *penetrance*, and the term *expressivity* denotes the manner of expression. Penetrance and expressivity are therefore phenomena that have confusing effects on the gene-character relationship. Both obviously mask the correspondence between genotype and phenotype. Incomplete penetrance and variable expressivity thus sometimes complicate the task of the plant breeder. Farmers, for the most part, have little liking for the variable unifoliate leaf phenotype of the Ventura variety, but many generations of selection against this dominant allele have been ineffective. Despite the unsightliness of its phenotypic effects, this allele is evidently not dysgenic.

## THRESHOLDS

The threshold effect is another phenomenon that weakens the gene-character relationship. Atlas, a traditional Mediterranean barley variety that is still widely grown, exhibits at least two temperature-sensitive threshold effects. One pale yellow/albino type has been observed only in seedlings that have been exposed to artificially induced near-freezing temperatures shortly after emergence. However, nearly all albino or yellow seedlings have small sectors of green in their leaves, and they may make a complete recovery when temperatures increase to ~20°C. Occasional Atlas barley seedlings also sporadically exhibit a different type of chlorophyll deficiency when they germinate in unusually cold and wet patches under field conditions; some mortality often results unless growing conditions improve rapidly. The concept of threshold effects appears to be useful in considering apparently genetically homogeneous populations made up of a genotype that occasionally behaves anomalously.

## ENVIRONMENTAL EFFECTS ON GENE EXPRESSION

The examples discussed in the previous sections of this chapter were deliberately chosen to show that the expressions of individual alleles can be conspicuously altered by factors of the genetic and/or external physical environment. If such genes were the norm in respect to instability under environmental stresses and/or changes in genetic background, selection and evaluation of performance could be very difficult to understand and/or to carry out. Fortunately, most genotypes express themselves predictably within the range of the conditions, genetic and environmental, they are likely to encounter. Most multilocus genotypes, particularly those that have evolved under a specified set of environmental conditions, are well buffered against common variations in their environment. Consequently, their performance in one or a very few sets of conditions provides a rather dependable guide to their performance under all except quite deviant environmental conditions. Nevertheless, the complex nature of the gene-character relationship leads to the expectation that performance may sometimes be altered because of deviant environmental and/or genetic circumstances. Plant breeders must therefore remain alert to truly dysgenic departures from previous performance in the genotypes they select from their hybrids.

## SEGREGATION AND RECOMBINATION IN HYBRIDS

Mendel deduced his two laws of heredity, laws that provide the basis for understanding particulate inheritance, from studies of single inherited differences. He worked with well-defined characters, such as yellow versus green in seeds, and did not consider joint behaviors until he understood individual behaviors. Once the 3:1 ratio was understood, interpretation of two-locus ratios such as 9:3:3:1 and three-locus ratios such as 27:9:9:9:3:3:3:1 followed much as a matter of course. This approach theoretically permits differences of any order of complexity to be analyzed, provided each allele has a uniquely recognizable effect. However, in practice gene-by-gene analyses of large numbers of loci have rarely been carried out, and plant breeding has not become merely a tedious exercise in assembling large numbers of favorable genes in ways that were predicted by some early geneticists. The first reason this has not happened is that most genes are not entirely independent in action; they interact in intricate ways and their joint and, particularly, their multiple-locus effects are often difficult to identify. A second reason has its basis in the spectacular rates at which potential for interactive complexity increases as the numbers of segregating alleles increase. This is illustrated in Table 7-1 in terms of numbers of allelic pairs.

The values in the columns of Table 7-1 are readily obtained by reasoning from Mendelian principles. The generalized predictions given in the bottom row of each column can be deduced from the simple situations. It is apparent from the general predictions that numerical complexity and potential multilocus interactive complexity increase exponentially with increases in the number of segregating allelic pairs. Both components in the general predictions—the base numbers and exponents—are positive integers. Consequently, an inevitable result of hybridization is an enormous increase in complexity when either the number of loci or the number of alleles per locus becomes greater. As an example of the latter, suppose there are

TABLE 7-1   Numerical characteristics of hybrids between parents differing in $N$ allelic pairs

| NUMBER OF ALLELIC PAIRS | KINDS OF 1 GAMETES POSSIBLE IN $F_1$ | KINDS OF 2 GAMETES POSSIBLE IN $F_2$ | SMALLEST PERFECT POPULATION IN $F_2$ | KINDS OF PHENOTYPES IN $F_2$ ASSUMING FULL DOMINANCE | KINDS OF PHENOTYPES IN $F_2$ ASSUMING NO EPISTASIS AND NO DOMINANCE |
|---|---|---|---|---|---|
| 1 | 2 | 3 | 4 | 2 | 3 |
| 2 | 4 | 9 | 16 | 4 | 9 |
| 3 | 8 | 27 | 64 | 6 | 27 |
| 4 | 16 | 81 | 256 | 16 | 81 |
| 10 | 1,024 | 59,049 | 1,084,576 | | |
| 21 | 2,097,152 | 10,460,353,203 | | | |
| $N$ | $2^N$ | $3^N$ | $4^N$ | $2^N$ | $3^N$ |

[1] Also gives numbers of genotypes in backcrosses and number of homozygous genotypes.

three instead of only two alleles at each of several loci. The number of kinds of genotype possible in families segregating for three alleles per locus increases at the much more rapid exponential rate of $6^N$, rather than the rate of $3^N$ as with two alleles per locus.

The consequences of hybridization, segregation, and recombination can be illustrated with a hypothetical cross between two homozygous lines of barley assumed to differ by a single allele of 14 loci, each locus located in a different arm of each of the 7 barley chromosomes; that is, the two homozygous lines differ by 14 independent (nonlinked) allelic pairs in total. Such an $F_1$ hybrid has the potential of producing $2^N = 2^{14} = 16,384$ different kinds of gametes that can be recombined to produce $3^N = 3^{14} = 4,782,969$ different genotypes in the $F_2$ generation. The smallest $F_2$ population that could be expected to contain, in expected Mendelian proportions, at least one of each kind of genotype would have to include $4^N = 4^{14} = 268,435,456$ plants, of which 16,384 would be homozygous at all 14 loci and $4^N - 2^N = 268,419,072$ would be heterozygous at 1 or more loci. Even if it were possible to grow and process an $F_2$ of the required size, the breeder would have no sure way of identifying homozygous genotypes, including the "best" genotype, in such a large population. Note, in particular, that nearly all genotypes in the $F_2$ (99.99%) would be heterozygous at least one locus and, hence, less prepotent than the 16,384 fully homozygous genotypes. Clearly, it would be advantageous to find some way of increasing the frequency of the "better" near-homozygous genotypes and at the same time decrease the frequency of the more heterozygous and less prepotent genotypes in barley, or any other self-pollinated species. The most widely used approach to this problem is to grow an $F_2$ population of practicable size (say 1,000 to 2,000 plants) and save selfed seeds from a number of desirable $F_2$ plants. These seeds can then be used to establish lines of descent within which plants can be allowed to self to increase homozygosity while selecting for favorable characteristics. There are various ways of calculating expected changes in homozygosity over generations within lines of descent; among these ways we will discuss only descent by selfing, because it is the simplest and the most widely used "line-breeding" scheme.

Starting with an $F_2$ plant heterozygous for $N$ gene pairs, the expected proportions of completely homozygous plants after $m$ generations of self-fertilizations are given by $[2^m - (1/2)^m]$. Hence, with 14 completely independent gene pairs, we expect after 3 generations of selfing that $[2^3 - 1/2^3]^{14} = [7/8]^{14} = 0.1542$, or about 15% of the plants, will be completely homozygous at all 14 loci. However, after an additional generation of selfing $[15/16]^{14} = 0.4051$, or about 40.5% of loci, are expected to be homozygous and one additional generation of selfing is expected to increase the percentage of completely homozygous loci to $[31/32]^{14} = 64.1\%$. Clearly, line breeding under selfing leads to dramatic increases in homozygosity and corresponding decreases in heterozygosity.

A more precise picture of the composition of selfed populations is given by expansion of the binomial $[1 + (2m - 1)]^N$ in which $m$ is again the number of generations of selfing and $N$ the number of gene pairs. In the expanded binomial the first exponent of each term gives the number of heterozygous loci, and the second exponent the number of homozygous loci. The arithmetic is tedious with large numbers of gene pairs. However, the main message emerges from simple cases;

hence, we will consider a population heterozygous in $F_1$ for only three gene pairs and selfed for four generations. Substituting into the binomial we obtain

$$[1 + (2^4 - 1)]^3 = [1 + 15]^3 = 1^3 + 3(1)^2(15) + 3(1)(15)^2 + 15^3.$$

Expressed in population terms, this indicates that our hypothetical $F_5$ is expected to include:

1 plant with 0 homozygous and 3 heterozygous loci
45 plants with 1 homozygous and 2 heterozygous loci
675 plants with 2 homozygous and 1 heterozygous loci
3,375 plants with 3 homozygous and 0 heterozygous loci

The ultimate aim of a barley breeder would normally be to obtain the single "best" homozygous line that any hybrid population is capable of producing. The effect of the selfing is a dramatic reduction in the frequency of the more heterozygous sorts and a corresponding increase in the frequencies of the more homozygous sorts (which are by far of greatest interest to a barley breeder). The foregoing calculations show that the $F_5$ generation is not only expected to be dominated by fully homozygous genotypes, but also to include some plants heterozygous at one or two loci. These heterozygous individuals are also likely to be interesting to the breeder because they are capable of producing in the next few generations (particularly selfed generations) full homozygotes that may not have appeared previously. Plant selections made in the $F_5$ generation are thus likely to be homozygous, or very nearly so, and the performances of their progeny are consequently much more predictable than the performance of the heterozygous genotypes that dominated earlier generations. For this reason, some breeders prefer to delay single-plant selection until later generations when prepotency is greater.

In practice, the calculated expectancies will not hold unless each genotype is equally productive. There is plausible experimental evidence that natural selection often favors alleles that enhance adaptedness and that human-directed selection may sometimes hasten the emergence of superior homozygous genotypes. Despite lack of precise information concerning the numbers of alleles involved, an important conclusion is inescapable: most hybridizations are capable of creating populations that are vastly complex genetically, especially in early-segregating generations. Another conclusion also seems inescapable, namely, that the numbers of allelic combinations possible from most hybrids are so large that only a small proportion can be dealt with effectively in a single cycle of hybridization. This suggests that populations should be as large as practicable in early generations, and that their management should be directed toward obtaining, in later generations, manageable numbers of promising pure lines for intensive evaluation of agricultural worth. Such evaluations, it is hoped, will reveal at least one pure line deemed good enough to be released to farmers. If not, at least some pure lines may be found that the breeder considers to be worthy of use as parents in additional cycles of hybridization.

## IMPROVING ELITE INBRED LINES OF CORN

We now turn to another problem that arises as a consequence of hybridization, segregation, and recombination. Corn was domesticated by early Central American Indians who, using mass selection methods, developed many cultivated open-pollinated races. Later many North American farmers and seedsmen also used mass selection to develop improved local varieties that, in turn, were the sources of the inbreds that led to successful hybrids. Modern open-pollinated varieties have thus survived hundreds of generations of selection in cultivation, and it is not surprising that such varieties have become the best genetic resources for developing inbred lines that enhance performance when hybridized. In developing new inbred lines, the choice of germplasm for breeding material is presently almost entirely limited to "elite" inbreds. Although thousands of new inbred lines are developed and tested annually throughout the world, only a few find wide use. Fewer, perhaps no more than 20, are in wide use worldwide at any given time, and still fewer survive over long periods of time. The development of an inbred line that achieves elite status is thus, without question, a rare event. Elite inbreds are so important that a great deal of corn breeding is concerned with conserving as much of the multilocus genotype of elite inbred lines as possible, while correcting deficiencies that keep elite lines from being even better. This brings us to another consequence of hybridization, segregation, and recombination. Crossing an elite inbred line to other materials, even another elite inbred line, leads to segregation and recombination in $F_2$ and later generations that nearly always results in severe loss of yielding ability in the $F_1$ hybrid progeny of the pure line. How can deficiencies that reduce the effectiveness of elite inbred lines be fixed or reduced while retaining the hard-won genetic virtues that elevated such lines to elite status? A common deficiency of corn, which was derived from a tropical grass, is lack of adaptedness to extended day length and low minimum temperatures. Consequently, a major activity in modern corn breeding is to make elite lines earlier in maturity to improve their adaptedness in short-season areas, particularly in the expanding corn-growing areas of northern latitudes. If a single highly heritable allele were known that would in itself cure the late flowering and/or other problems of adaptedness in corn, a standard backcross breeding program might solve the problem quickly and effectively. Recurrent backcrossing to an elite homozygous inbred line, with selection each generation for the desired allele, would be expected, within five or six generations, to produce an early-flowering version of the formerly late-flowering elite inbred line. Unfortunately, few if any such major alleles for earliness (or related traits affecting earliness) are known. Instead, such traits behave as if they are governed by many loci, each with relatively small effects on earliness. The plans that have most often been adopted to solve this difficult problem usually start by crossing one early elite inbred line to an elite late inbred line and immediately crossing each $F_1$ hybrid to its late elite inbred parent. A large first backcross (BC1) population is grown, and BC1 plants that are early (and otherwise reasonably desirable) are selfed. These $S_1$ progenies are expected to be fully homozygous at $[2^m - 1/2^m)^N$ loci, where $m$ is the number of generations of selfing and $N$ is the number of gene pairs. Assuming 10 gene pairs governing earliness, the $S_1$ progenies are expected to be fully homozygous at $0.50^{10} = 0.0010 = 0.1\%$ of their loci. The poorer $S_1$ progeny rows are discarded, the better plants in the retained rows are again selfed, and their $S_2$ progeny are grown ear to row; about 5% of these $S_2$ plants are expected to be fully homozygous at $0.75^{10}$, or ~6% of their loci, and 94% are expected to be homozygous at some loci and heterozygous at other loci. At this point some breeders may choose to cross the best early plants in the best rows to the original late-flowering elite parental inbred line to initiate a second cycle. Other breeders may choose to self again and select the best plants in the better $S_3$ rows, among which a higher proportion would be expected to be homozygous at all 10 loci. After perhaps one or two additional cycles of backcrossing and selfing, when breeders feel that they may have accomplished their main purpose (transfer of earliness alleles into the genotype of the late inbred line), they can bulk seeds of the best lines and make crosses with an array of early-flowering elite inbred lines to obtain seed for testing the specific combining ability with elite early lines of the modified lines. Many variations of this basic plan have been used to develop useful early inbred lines of corn. Few breeders use precisely the same scheme, nor is any one breeder likely to use the same scheme in exactly the same way for different source materials.

## LINKAGE

In examining the consequences of hybridization the effects of linkages were avoided by carefully restricting hypothetical examples to situations in which linkage would not likely be involved (all loci in different arms of chromosomes), or the likely effects were vastly complicated and hence deliberately ignored (the corn example). In actuality, however, most genotypes are known to carry different alleles at many loci in different chromosome arms; consequently, linkages are expected to be common-place in hybrids, and the breeder should routinely take possible confusing effects of linkage on recombination into account. With independent inheritance, assortment is free and all combinations are equally frequent. Linkage has the general effect of causing an overabundance of parental combinations and a corresponding deficiency in recombinants. The magnitude of this effect is shown in Table 7-2, in which the recovery of parental and recombinant types is given for the two-locus case for several linkage intensities. Numerical outcomes become quite complicated when three loci are linked and very complex when the segregations of several linked loci, each with several alleles, are considered simultaneously.

Generally speaking, modern breeders work with materials that have already been altered extensively; thus, linkage can usually be regarded as a conservative influence, tending to hold existing favorable combinations of genes together. In recently domesticated species or in populations that have been under improvement for only short periods in a given environment, it is unlikely that desirable alleles will be concentrated in any single multilocus genotype at most loci. Under such circumstances tight linkage may be a hindrance to relatively free recombination, and this may inhibit assembly of new favorable combinations. Conversely, highly adapted genotypes are likely to have favorable alleles at many loci. Thus, linkage can aid the plant breeder in such situations because it tends to hold existing favorable combinations together.

TABLE 7-2  The effect of linkage on the proportion of AB/AB genotypes expected in $F_2$ from backcrosses to the double heterozygote

| | PERCENTAGE OF AB/AB INDIVIDUALS IN $F_2$: IF THE $F_1$ IS: | |
| RECOMBINATION VALUE | AB/ab | AB/ab |
|---|---|---|
| 0.50 (independence) | 6.25 | 6.25 |
| 0.25 | 14.06 | 1.56 |
| 0.10 | 20.25 | 0.25 |
| 0.02 | 24.01 | 0.01 |
| 0.01 | 24.50 | 0.0025 |
| $p$ | $1/4\,(1-p)^2$ | $1/4p^2$ |

Table 7-2 shows that tight linkages between favorable and undesirable alleles can drastically delay the progress of breeding programs, and many examples of this situation have been encountered. Alleles that confer resistance to pathogenic diseases, particularly alleles extracted from wild populations, often have associated undesirable effects (e.g., late maturity, inferior product quality, negative effects on yield in cultivation), which have been difficult to disassociate from disease resistance. In some cases Herculean efforts have failed to break such associations, which suggests that the resistance allele may produce some product that is bad for the pathogen and also bad for the cultivated host. Conversely, linkages between favorable alleles can aid breeding programs. In the broad view, each step forward in plant breeding builds up the number of favorable associations, and whenever the favorable associations are between alleles of linked loci, the association increases the. value of the linked group of alleles as parents in future breeding endeavors. Thus, breeders tend to use the best available varieties as parents and avoid "primitive" unimproved materials as parents.

# EIGHT

# *Inheritance of Continuously Varying Characters: Biometrical Genetics*

Mendel's choice of several different characters, each unambiguously classifiable into alternative readily identifiable (either/or) phenotypic classes, was the key to his success in deducing a posteriori the particulate nature of heredity. Mendel was mindful of the many gradations that occurred within some of his classes (e.g., the short and tall classes of peas), but he disregarded such variability probably because he realized, apparently intuitively, that it would not be easy to explain in terms of the simple hereditary particles on which he had focused his attention. Darwin considered small, continuous heritable changes to be the basic steps in evolution. The biometrician Galton, as well as many plant and animal breeders of the nineteenth century, were also aware that continuously varying characters were at least partly heritable. However, as discussed in Chapter 6, it remained for Johannsen, Nilsson-Ehle, and East to provide, early in the twentieth century, convincing experimental evidence that alleles of Mendel's particulate "factors" or "elements," now called *genes*, were responsible not only for discretely inherited characters but also for continuously varying characters.

Most of the characters plant breeders wish to improve vary continuously. The genetics of such continuously varying characters understandably soon came to be called *quantitative* or *biometrical genetics*. The purpose of this chapter is to outline the early ways in which the ideas of Johannsen, Nilsson-Ehle, and East were elaborated and incorporated into Mendelian theory. In dealing with quantitative characters the study of discretely different kinds of individuals is precluded by the very nature of the data, and it is apparent that observations entirely different in kind from those of Mendel would be needed to obtain the type of genetic information that allows prediction of genetic advance when selection is practiced under various breeding plans. This need was recognized by a number of early twentieth-century geneticists, particularly R. A. Fisher (England), who developed genetic models a priori from Mendelian theory but whose observations appropriately took an entirely different form—means, variances, and other statistics that can be calculated from measurement data. Such statistics allow estimates of numerical attributes (parameters) of the genotype-environmental milieu that influence breeding behavior. In this approach the by then familiar parameters of Mendelian genetics were abandoned and genetic significance was attached to different parameters. This in no way means that the approach was non-Mendelian. It merely recognized the necessity for specifying the effects of genes affecting quantitative differences in ways that were regarded as appropriate to the task. More recently it has been found that linkages between alleles of quantitative characters and discretely recognizable Mendelian loci can be useful in analyzing quantitative characters. The utility of such linkages in analyzing quantitative characters and in plant breeding are considered and in later chapters, especially Chapters 10 and 11.

## QUALITATIVE VERSUS QUANTITATIVE CHARACTERS

That both continuous and sharply discontinuous variation are observed in many characters establishes that the distinction between qualitative and quantitative characters is not clear-cut. Seed size, as in Johannsen's beans, is usually a quantitative character governed by many loci of small effect. So also is stature, but short and tall strains conditioned by single-gene differences have been found in all or nearly all plant species in which a serious search has been made. Nilsson-Ehle's wheat and oats suggested an intermediate situation: three major loci, each with modest effect, appeared to be responsible for continuous variation in seed color in these plants. In practice, then, the difference between qualitative and quantitative characters depends not only on the number of genetic loci involved but also on the relative importance of heredity and environment in determining the expression of phenotypic characteristics. It therefore became apparent that the key to progress in the analysis of quantitative characters lay in evaluating the relative contributions of heredity and environment in their effects on observed phenotypic variability.

## THE CONCEPT OF HERITABILITY

In a strict sense, the question of whether a characteristic is a result of heredity or environment is meaningless. Genes cannot cause a character to develop unless the organism is growing in an appropriate environment, and, conversely, no amount of manipulation will cause a phenotype to develop unless the necessary gene or genes are present. Nevertheless, it came to be recognized that the variability observed in some characters might result primarily from differences in the numbers and the magnitude of effect of different genes, but that variability in other characters might stem primarily from the differences in the environments to which various individuals had been exposed. It therefore appeared essential to have reliable measures of the relative importance of not only the numbers and magnitudes of the effects of the genes involved, but also of the effects of differing environments in determining the expression of phenotypic characters.

In developing such a quantitative statement, it is convenient to start with the simplest possible monogenic inheritance and to vary the environmental variability as a proportion of the total. In Figure 8-1($A$), a 12-unit difference between parents ($P_1 = 50$ units and $P_2 = 62$ units) postulated to be due entirely to a single Mendelian gene pair. Assuming no dominance, the genetic variance $\sigma_G^2$ (the mean squared

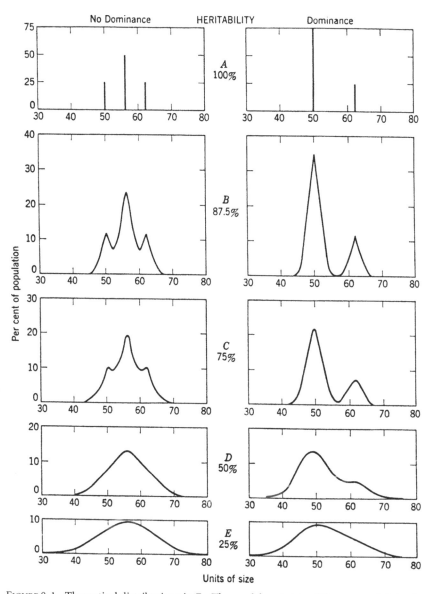

FIGURE 8-1  Theoretical distributions in $F_2$. The model postulates (1) monogenic inheritance, (2) a twelve-unit difference between the parents, and (3) that the effect of environment varies from nil (100 per cent heritability) to the point where environmental effects account for three-fourths of the total variability (25 per cent heritability). The left column depicts no dominance; the right column, full dominance. Note that the scale for A differs from the scale for B through E.

deviation from the mean) can be calculated to be $1(6)^2 + 2(0)^2 + 1(6)^2/4 = 18$; assuming full dominance, the genetic variance is $3(3)^2 + (9)^2/4 = 27$.

Because the genotype completely determines the phenotype in Figure 8-1(A), the heritability is 100% and each phenotype can be represented by histograms, as shown in this figure. If, however, environment also contributes to the variability in a measurable degree, the $F_2$ genotypes can no longer be represented by limited numbers of histograms but can more conveniently be portrayed by curves. The several curves of Figure 8-1($B$) were constructed on the assumption that one-eighth of the total variability, as measured by variance ($\sigma^2$), is due to environment and seven-eighths is genetic in origin; that is, that the heritability is 87.5%. Clearly, heritability ($H$) specifies that portion of the total variability that results from genetic causes, that is, the ratio of genetic variance to total variance. Expressed quantitatively, $H = \sigma_G^2 / \sigma_G^2 + \sigma_E^2$, in which $\sigma_G^2$ and $\sigma_E^2$ are the genetic and environmental variances, respectively.

Note that the curves representing the three possible genotypes in Figure 8-1($A$) and ($B$) are distinct only at the two higher heritabilities. In Figure 8-1($C$) (75% heritability) the curves overlap to a small extent. Note also that the degree of overlap increases progressively with still lower heritabilities, with the result that many homozygous large plants as well as many homozygous small plants can no longer be distinguished from heterozygotes, even though genetic control is monogenic. If, however, selection were limited to only the largest large plants and the smallest small plants, the likelihood of selecting homozygous large or homozygous small plants remains good down to $H \sim 25\%$. Had larger differences been postulated between the original parents, discreteness would have been maintained with lower heritabilities. Note also that overall curves for the $F_2$ generation, obtained by summing the curves for each of the three genotypes, approach normality more and more closely as heritabilities decrease. Thus, even though dominance leads to conspicuous skewness when only one gene pair is postulated, skewness is hardly detectable with 25% heritability. It can be anticipated that in real situations, dominance might lead to substantial problems in selection should heritabilities be low. For example, heritability for yielding ability, when measured in actual experiments, is often not significantly different from zero, and experimental estimates of heritability for corolla length in *Nicotiana*, selected by East because of the remarkable insensitivity of this character to environmental influences, is usually on the order of 75% (the value of $H$ assumed in constructing the curves of Figure 8-1($C$)).

We turn now to Figure 8-2, which illustrates the expected effects of varying the genetic situation from one to two, then from four to six, then from six to twelve equally effective gene pairs, all the while holding heritability constant at 100%. With no dominance (left column) the distributions remain symmetrical. It can be seen from Figure 8-2 (left column) that even with as few as four to six more or less equally effective gene pairs, these distributions might be difficult to distinguish from normal bell-shaped distributions, even should the environment have very little or even no effect. The right column of Figure 8-2 also shows how skewness, even if due exclusively to isodirectional dominance (dominance in the same direction for all loci), becomes increasingly difficult to detect as numbers of loci increase. When modest environmental effects are incorporated into the genetic models of Figure 8-2, the distributions become difficult to distinguish from normal symmetrical distributions.

The genetic models on which Figure 8-2 are based make the highly unlikely assumption that all gene pairs have equal effects on the phenotype, that dominance is isodirectional, that allelic frequencies are 0.5 at all loci, and that there are no

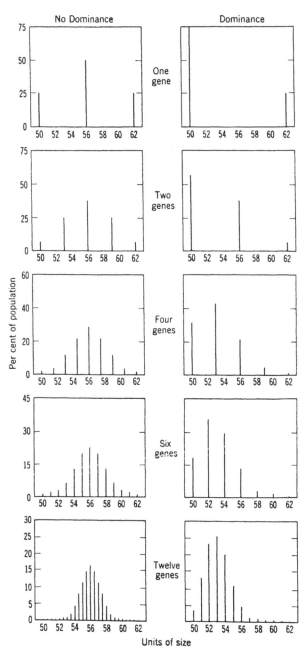

FIGURE 8-2    Theroetical distributions in $F_2$. the model postulates (1) 100 per cent heritability, (2) that a twelve-unit difference between the parents is governed by various numbers of genes of equal effects, (3) no linkage, and (4) that dominance is isodirectional. The
scale is not constant.

linkages. If all of the gene pairs do not have the same effect, the distributions tend to resemble those produced by fewer gene pairs. The same effect is produced by coupling-phase linkages. Zygotes for each locus are expected to be distributed according to $[p(A) + q(a)]^2$. If $p \neq q$, the variance will be smaller; that is, the population will be less variable than if $p = q = 0.5$ at all loci. A dampening effect on variability also results from repulsion-phase linkages. Despite such complications, Figures 8-1 and 8-2 make it clear that either moderately low heritabilities and/or moderately large numbers of gene pairs will usually result in continuous and more or less normal distributions.

## FACTORS AFFECTING PHENOTYPIC EXPRESSION

With Johannsen's delineation of the distinction between genotype and phenotype and the experimental confirmations by East and Nilsson-Ehle that quantitative characters comply with Mendel's rules, it became clear that quantitative variation results from the joint actions of genotype and environment. In quantifying this joint action, the convention usually adopted has been to express phenotypic values in linear fashion so that the expression of any given gene affecting a quantitative trait, denoted by A, is represented as

$$A = \mu + a + e + ae. \tag{8-1}$$

In Equation 8-1, the numerical value $A = \mu + a + e + (ae)$ is expressed as the sum of a general population mean ($\mu$), a genotypic effect ($a$), an environmental effect ($e$), and a multiplicative effect of genotype by environmental interactions ($ae$). Were it possible to characterize all genotypes in all environments, the value obtained would represent the general population mean. Any single genotype in any single environment adds or subtracts from the mean, depending on whether its effect on character expression falls above or below the general population mean. The interaction term will be zero only if all genotypes behave the same in all environments; that is, that there are no genotype x environment interactions. All measurements are necessarily on phenotypes, so that any practical definition of $\mu$, $a$, $e$, and $ae$ must also be in phenotypic terms. Plant breeders normally restrict attention to environmental factors that are likely to be important in determining the agricultural value of genotypes which, in practice, are often identified with geographical locations ($L$) and years ($Y$) within some comparatively homogeneous geographical area. Hence, after the first-order effects ($L$ and $Y$) have been taken out, hierarchies of first-order ($VL$, $VY$), second-order, and higher-order interactions can also be isolated and tested for significance, usually against "error" variance calculated from replicates within lower-order factors. If the analysis is conducted in terms of different locations and years for a given set of genotypes, Equation 8-1 expands to

$$A = \mu + a + r + 1 + (al) + (ly) + (aly) + e \tag{8-2}$$

In this expression, $\mu$ and $a$ remain the population mean and the genotypic effects, respectively; $r$, $l$, and $y$ are the direct effects of replicates, locations, and years, respectively; $al$, $ly$, and $aly$ are interaction effects; and $e$ is a composite error term of remaining effects (including, for example, differences from plant to plant in the same replication, differences among plants in the same family, and errors of measurement). Equation 8-2 therefore identifies the principal factors of the environment that bear on phenotypic expression but does so in terms of locations and years, without attempting to specify in any exact way environmental influences that cause genotypes to vary relative to one another within a usually quite homogeneous region. Other factors affecting phenotypic expression can be added—planting dates, irrigation versus no irrigation, fertilizer versus no fertilizer—and in suitably designed experiments, it is possible to extract estimates of additional genetic, environmental, and interaction variances. It should be noted that interaction effects are especially troublesome because they are multiplicative and hence tend to have large errors of estimation. It is easily seen that complex experiments typically lead to troublesome complexities in estimation and interpretation so that fewer and fewer breeders are willing to attempt such experiments. Even apparently simple experiments often reveal tangled complexities that can be resolved only by experimentation, which often overwhelms the resources of even the best-supported breeders. Despite complexities that severely limit their usefulness, experiments to estimate components of genetic and environmental variance have heuristic value, and we will consequently develop the theory and illustrate difficulties of application with examples.

## GENETIC ADVANCE UNDER SELECTION

The essential features of most breeding programs are (1) selection within a base population of genetically variable individuals and (2) utilization of the materials selected over several generations, either as potential commercial varieties or as parents in crosses to be subjected to additional cycles of selection. It is apparent from Equations 8-1 and 8-2 that genetic advance under selection, as represented by improvement in $a$ (a genotypic value) in the selected population, will depend on (1) the amount of genetic variability (the magnitude of the differences in $a$ among different individuals in the base population) and (2) the magnitude of the effects of environmental ($e$) and interaction components ($ae$) of variability in masking the genetic effects ($a$). A third factor, the intensity of selection that is practiced, will obviously also affect the rate of genetic advance under selection.

Starting with the simplest case, selection among a large number of fully homozygous lines descended from fully homozygous individuals of a self-pollinating population, mean yield (denoted by $A$) can be calculated from measurements made on several individuals of each line. Assuming that the values of $A$ are normally distributed with standard deviation, $\sigma^2 A$, the frequency distribution of the mean values can be represented as shown in Figure 8-3. If a portion $q$ of the total of $N$ lines, say all lines with phenotype $A$ greater than some fixed value, $A$, are selected, the selected lines $\bar{a}_s$ will be those falling in the shaded area of Figure 8-3. It can be shown that the expected genetic advance for the selected families will be

$$G_s = (k)\,(\sigma^2 A)\,(H). \tag{8-3}$$

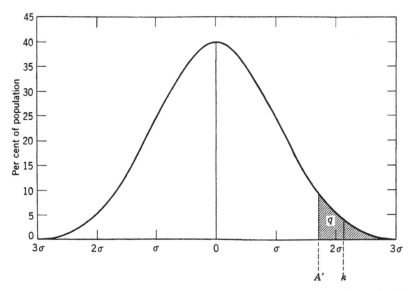

FIGURE 8-3   Theoretical distruction of mean yields of lines compared in replicated-yield trials. When the top 5 per cent of the line are selected, that is, those with yields larger than $A'$ ($= 1.65\sigma$), the mean of the $q$ selected families is expected to be $k$ (in standard measure) and $k\sigma$ when expressed in terms of the actual units in which the measurement were taken.

The symbols in Equation 8-3 have the following significance: $G_s$ (expected genetic advance) represents the difference between the mean genotypic value ($\bar{a}_s$) of the $q$ selected lines and the mean genotypic value of the $N$ original lines ($\bar{a}$); that is, $G_s = \bar{a}_s - \bar{a}$. $\sigma^2 A$ is the phenotypic standard deviation of the mean yields of the $N$ original lines. $H$ is the coefficient of heritability, estimated as the ratio of genotypic to phenotypic variance, $H = \sigma_a^2/\sigma_A^2$; $k$ is a selection differential that takes into account the mean phenotypic value of the $q$ selected lines ($A'_s$), the mean phenotypic value of all $N$ lines tested, the phenotypic standard deviation $\sigma_A^2$, and the stringency of selection $q/n$. Because $k$ is expressed in standard deviation units, that is, in terms of a unit curve, $k$ varies only as selection intensity, $q/n$, varies. Thus, if the highest 1% of lines are saved, $k$ takes value 0 2.64. Other likely selection intensities for plant materials are 2, 5, 10, 20, and 30% of lines saved for which selection intensities, $k$, take values of 2.42, 2.06, 1.76, 1.40, and 1.16, respectively.

If heritability were 100% (i.e., if $\sigma_A^2 = \sigma_a^2$), the phenotype would provide an exact measure of the genotypic value. But even in this unlikely event the heritability value alone provides an inadequate measure of genetic advance, because both $k$ and $\sigma_A^2$ also enter the prediction equation. Clearly, for given heritability, greater genetic progress is expected when $k$ and/or $\sigma_A^2$ take large, rather than small, values. Equation 8-3 is thus seen to evaluate, relative to each other, each of the three factors (amount of genetic variability, heritability, and intensity of selection) that are involved in genetic advance under selection.

## GENETIC COMPONENTS OF CONTINUOUS VARIATION

When families are selected from a base population and used to form a filial population, the net expected improvement in the next generation is given by Equation 8-3. This method of prediction works best with base populations that are made up of homozygous pure lines (often obtained by selfing the base population for several generations) because such materials reproduce themselves precisely, aside from deviations that might result from rare mutations and/or from outcrosses between different homozygous pure lines that occur during the experiment. Additional genetic factors must be taken into account when the materials subjected to selection reproduce themselves less precisely than homozygous lines. The most likely problems result from the segregation and recombination that may take place in following filial generations or from selection that alters the frequency of different emergent homozygous lines.

Partition of hereditary variance into component parts was first performed by Fisher (1918). He recognized three components of hereditary variability (Table 8-1): (1) an additive portion stemming from the difference between homozygotes

TABLE 8-1   Gene models for two loci each with two alleles

| Model I, Additive | | | | Model II, Dominance | | | |
|---|---|---|---|---|---|---|---|
| *AABB* | *AABb* | *AAbb* | *AA--* | *AABB* | *AABb* | *AAbb* | *AA--* |
| 7 | 6 | 5 | 6 | 4 | 4 | 2 | $3\frac{1}{2}$ |
| *AaBB* | *AaBb* | *Aabb* | *Aa--* | *AaBB* | *AaBb* | *Aabb* | *Aa--* |
| 5 | 4 | 3 | 4 | 4 | 4 | 2 | $3\frac{1}{2}$ |
| *aaBB* | *aaBb* | *aabb* | *aa--* | *aaBB* | *aaBb* | *aabb* | *aa--* |
| 3 | 2 | 1 | 2 | 3 | 3 | 1 | $2\frac{1}{2}$ |
| *--BB* | *--Bb* | *--bb* | | *--BB* | *--Bb* | *--bb* | |
| 5 | 4 | 3 | | $3\frac{3}{4}$ | $3\frac{3}{4}$ | $1\frac{3}{4}$ | |

| Model III, Complementary | | | | Model IV, Complex | | | |
|---|---|---|---|---|---|---|---|
| *AABB* | *AABb* | *AAbb* | *AA--* | *AABB* | *AABb* | *AAbb* | *AA--* |
| 3 | 3 | 1 | $2\frac{1}{2}$ | 4 | 2 | 3 | $2\frac{3}{4}$ |
| *AaBB* | *AaBb* | *Aabb* | *Aa--* | *AaBB* | *AaBb* | *Aabb* | *Aa--* |
| 3 | 3 | 1 | $2\frac{1}{2}$ | 4 | 3 | 1 | $2\frac{3}{4}$ |
| *aaBB* | *aaBb* | *aabb* | *aa--* | *aaBB* | *aaBb* | *aabb* | *aa--* |
| 1 | 1 | 1 | 1 | 3 | 2 | 1 | 2 |
| *--BB* | *--Bb* | *--bb* | | *--BB* | *--Bb* | *--bb* | |
| $2\frac{1}{2}$ | $2\frac{1}{2}$ | 1 | | $3\frac{3}{4}$ | $3\frac{1}{2}$ | $1\frac{1}{2}$ | |

Numbers indicate the genotypic value, $\mu + a$, for each genotype. The border rows and columns represent the mean genotypic values for the three phases possible at each locus, assuming gene frequency is 1/2 at each locus.

at single loci, (2) a dominance component arising from interactions among alleles of the same locus (intra-allelic interactions), and (3) interactions among alleles of different loci (interlocus interactions of nonalleles, epistasis). This classification has often been used in describing types of gene action and in estimating the importance of various types of gene actions and interactions. Table 8-1 includes four models that illustrate the effects of these types of gene action and the interactions among those types that are perhaps most likely to be encountered.

The simplest model of quantitative gene action postulates that the substitution of one allele for another leads to a plus or minus shift on the scale of measurement and that these shifts remain the same regardless of other genes that may be present. This situation is illustrated in the additive model of Table 8-1. The effect of replacing *a* by *A* is two units, and that of replacing *b* by *B* is one unit. There is no dominance, because the effect is the same in homozygotes as in heterozygotes. There is also no epistasis, because the effect remains the same regardless of phase at the other locus. This type of gene action in pure form produces a 1:2:1:2:4:2:1:2:1 segregation ratio in $F_2$ populations. The dominance model of Table 8-1 illustrates complete dominance without epistasis. Alleles at a single locus interact in such a way that substitution of *A* for *a* in genotype *aa* is not the same as in genotype *Aa*. This type of gene action (full dominance) in pure form produces a ratio of 9:3:3:1 in $F_2$. If all loci acted in this way, there would be only two types of hereditary variance: additive effects and deviations from additive effects resulting from dominance. Table 8-1 also illustrates one of the classical types of epistasis, complementary gene action. The hereditary variance includes additive, dominance, and epistatic components. The fourth model of Table 8-1 illustrates complex interactions of the type that may be expected in many real situations. The three phases of the *A* locus illustrate full dominance, overdominance, and full recessiveness with *BB*, *Bb*, and *bb*, respectively, whereas the three phases of the *B* locus illustrate underdominance, partial dominance, and no dominance with *AA*, *Aa*, and *aa*, respectively.

In the description of genetic variability, hereditary parameters can be defined precisely in terms of Mendelian models. In a model with only two gene pairs, each with two alleles, the simplest on which interlocus interaction can be represented, nine genotypes are possible. The Mendelian variation in the system can be specified by eight parameters representing the eight independent comparisons among the nine genotypes. One of these parameters, $d_a$, in the notation of Hayman and Mather (1955), represents the difference between the homozygotes at the *Aa* locus (Table 8-1). Another parameter, $d_b$, represents the corresponding additive effect of the *Bb* locus. A third parameter, $h_a$, represents deviation of the heterozygote, *Aa*, from the *AA-aa* midparent—that is, the dominance effect of the *Aa* locus; $h_b$ represents the corresponding deviation of the *Bb* heterozygote. The four remaining degrees of freedom represent interactions among the two additive and two dominance effects.

The contributions to genetic variability in the $F_2$ generation for the four models are given in Table 8-2. The total genetic variance in an $F_2$ with completely additive gene action (Model 1) can be calculated as $\frac{1}{16}(7)^2 + \frac{1}{8}(6)^2 + \frac{1}{16}(1)^2 - 4^2 = 2\ 1/2$. As shown in Table 8-2, two units of the total are due to $d_a$ and 1/2 unit to $d_b$. Thus, 100% of the genetic variability is additive, and all additive effects are attributable to differences between the *AA-aa* and *BB-bb* homozygotes. The total genetic

TABLE 8-2 Genetic variances in $F_2$ populations

| HEREDITARY PARAMETER | ADDITIVE MODEL | DOMINANCE MODEL | COMPLEMENTARY MODEL | COMPLEX MODEL |
|---|---|---|---|---|
| $d_a$ | 2 | $\frac{2}{16}$ | $\frac{18}{64}$ | $\frac{18}{256}$ |
| $d_b$ | $\frac{1}{2}$ | $\frac{8}{16}$ | $\frac{18}{64}$ | $\frac{9}{256}$ |
| $h_a$ | 0 | $\frac{1}{16}$ | $\frac{9}{64}$ | $\frac{9}{256}$ |
| $h_b$ | 0 | $\frac{4}{16}$ | $\frac{9}{64}$ | $\frac{9}{256}$ |
| $d_a d_b$ | 0 | 0 | $\frac{4}{64}$ | $\frac{4}{256}$ |
| $d_a h_b$ | 0 | 0 | $\frac{2}{64}$ | $\frac{4}{256}$ |
| $d_b h_a$ | 0 | 0 | $\frac{2}{64}$ | $\frac{18}{256}$ |
| $h_a h_b$ | 0 | 0 | $\frac{1}{64}$ | $\frac{25}{256}$ |
| Total variance | $2\frac{1}{2}$ | $\frac{15}{16}$ | $\frac{63}{64}$ | $\frac{255}{256}$ |

variance for the dominance model in $F_2$ is only about 1/3 as large as that of the additive model, or, more precisely, $15/16 \div 2.5 = 0.9375 \div 2.5 = .375$ (Table 8-2). Thus, 10/16 of the total variance is attributable to the differences between the *AA-aa* and *BB-bb* homozygotes; the remainder, 5/16, arises from failure of the $A_a$ and $B_b$ heterozygotes to fall on the *AA-aa* and *BB-bb* midparental values, respectively. The total genetic variance for the complementary model is still smaller, 63/64 (Table 8-2), of which approximately 57% is attributable to additive, 29% to dominance, and 14% to epistatic effects of the two loci concerned. For the complex model the total genetic variance is 255/256 (Table 8-5), of which additive, dominance, and epistatic effects are in the ratio of 70:4:26, respectively. These models illustrate that in interacting systems involving dominance and epistasis, genes can still be described in terms of their average effect in generation $F_2$ and a portion of the variance can be ascribed to them, even though they have lost nearly all of their singularity of effect. Expanding models to three or four loci leads to further large increases in the complexity of genetic interactions of higher orders, with the result that interpretations of genetic components almost certainly require still larger numbers of seriously debilitating assumptions. The most frequent among such assumptions is that not only higher-order interactions but even second-order genetic parameters such as $d_a d_b$, $d_a h_b$, $d_b h_a$ and $h_a h_b$ are assumed to take values of zero. Keeping these substantial difficulties in mind, we now turn to an apparently simple experiment designed to test whether experiments to estimate components of variance are likely to be useful in making practical decisions concerning breeding practices.

## AN EXPERIMENT TO ESTIMATE HEREDITARY PARAMETERS

Considering only one locus of our model, the values associated with the three possible genotypes are $AA = d_a$, $Aa = h_a$, and $aa = -d_a$ (Figure 8-4). All deviations are measured from the *AA-aa* midparent; consequently, the difference between the *Aa* and *aa* homozygotes is $2d_a$. The capital-letter symbol refers to the allele with the

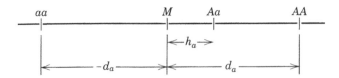

FIGURE 8-4   Values associated with three possible genotypes.

plus effect and denotes neither dominance nor recessiveness. When a character is governed by a number of genes, the phenotype of a true-breeding strain will be $\Sigma(-d) + c$, in which $c$ represents a constant base level dependent on genes not segregating as well as on nonheritable influences. Consequently, $c$ is typically not included in the estimation equation. Thus, the $F_1$ of a hybrid between *Aa* and *aa* will be phenotypically $h_a$. If $h_a = 0$, no dominance is indicated, whereas $h_a = d_a$ represents full dominance and $h_a > d_a$ (which has been found to be unlikely) represents overdominance. The $F_1$ between two homozygous strains differing in a number of genes is $\Sigma(h)$, in which $h$ can take sign. If an $F_1$ hybrid deviates from the midparent, some dominance obviously must exist at one or more loci. But the observation $\Sigma(h) = 0$ does not necessarily indicate that individual values of $h$ are zero for all loci, because individual dominance effects may have different signs and may counterbalance one another to at least some extent. When all (+) alleles by which two parents differ are concentrated in one parent and all (−) alleles in the other, the difference between strains must be $2\Sigma d$. In practice, however, isodirectional distribution of (+) and (−) alleles and equality of effect both seem highly unlikely; consequently, the true difference will usually be $< 2\Sigma d$ but $> 0$.

The constitution of the $F_2$ generation for any single gene *Aa* is expected to be $1/4AA$ (phenotype $d_a$): $1/2Aa$ (phenotype $h_a$): $1/4\ aa$ (phenotype $-d_a$), whereas the mean of the $F_2$ generation is expected to be $1/4 (+d_a) + 1/4 (-d_a) + 1/2 (h_a) = 1/2_{ha}$. The contribution of this gene to the sum of squares is $1/4 (d_a)^2 + 1/4 (-d_a)^2 + 1/2 (h_a)^2 = 1/2 (d_a)^2 + 1/2 (h_a)^2$, and, because the mean of generation $F_2$ is $1/2 h_a$, the variance of the $F_2$ is $1/2 (d_a)^2 + 1/2 (h_a)^2 - 1/4(h_a2) = 1/2(d_a)^2 + 1/4(h_a)^2$. Because a unit population is assumed, and equality of effect of all loci is also assumed, this is also the mean of the squared deviations from the mean, or the variance of the $F_2$ population. This is only the genetic portion of the variance; an environmental portion will almost certainly be found in real populations. Letting $D = \Sigma (d)^2$, $H = \Sigma(h)^2$, and letting $E$ denote the variance due to environmental causes, the variance of an $F_2$ segregating for several genes can be represented as

$$V_{F_2} = 1/2\ D + 1/4H + E \qquad (8\text{-}4)$$

By parallel methods, the summed variances of the backcrosses to the parents can be shown to be

$$V_{B_1} + V_{B_2} = 1/2D + 1/2H + 2E \qquad (8\text{-}5)$$

Equations for generations other than $F_2$, $B_1$, and $B_2$ can be derived in similar fashion.

An apparently very simple experiment serves to illustrate the many problems associated with the estimation procedure. In this experiment, heading date (date of emergence of the first spike from the boot for each plant) was determined for the wheat varieties Ramona ($P_1$) and Baart ($P_2$), for the $F_1$ and $F_2$, and for the backcrosses ($B_1$ and $B_2$) of the $F_1$ to $P_1$ and to $P_2$, respectively. The numbers of individuals scored and the means and variances for these generations are given in Table 8-3. It can be seen from this table that the means of the hybrid generations ($F_1, F_2, B_1, B_2$) fall between those of $P_1$ and $P_2$, but are not exactly intermediate. Table 8-3 also shows that in two generations ($F_1, B_1$) there is slight dominance in the direction of earliness, but in other generations ($F_2, B_2$) there is slight dominance in the direction of lateness. The $F_2$, $B_1$ and $B_2$ generations are much more variable than the parents or the $F_1$; hence, segregation is obvious and clear. Note also that the three estimates of $V_E$ ($P_1, P_2, F_1$) are rather variable and, in particular, that the $F_1$ appears to be substantially less variable than either parent. Note further that $B_1$, the backcross to the early parent appears to be much less variable than $B_2$, the backcross to the late parent, which suggests substantial dominance in the direction of earliness.

Using the mean of the variances (Table 8-3) of the three nonsegregating generations ($P_1$ = parent 1, $P_2$ = parent 2, and $F_1 = P_1 \times P_2$) as the single best estimate of $E(V_p + V_{F_1} + V_{F_2} = 11.036 + 5.237 + 10.320 = 26.593/3 = 8.894$) becomes the best estimate of $E$. Thus, in the $F_2$ generation the environmental variance $V_E = 8.864/40.350 = 0.22$, indicating that 22% of the total variance of the $F_2$ is environmental in origin and the remainder (78%) is due to genetic causes. Parallel estimates from the $B_1$ data give $8.864/17.352 = 51\%$ environmental and 49% genetic, whereas estimates from the $B_2$ data give $8.864/34.288 = 26\%$ environmental and 74% due to genetic causes. We are, however, more interested in heritability computed on the basis of the additive genetic variance because this estimate should better indicate the extent to which the progeny of the $F_2$, $B_1$, and $B_2$ plants resemble their parents. Note that the $F_2$, $B_1$, and $B_2$ data alone allow estimation of only total variance. However, simultaneous solution of the equations for $F_2$ and for $B_1 + B_2$ allows estimation of the additive effects of the genes involved, as follows:

$$2V_{F_2} = D + 1/2\ H + 2E = 80.700$$

$$V_{B_1} + V_{B_2} = 1/2D + 1/2H + 2E = 51.640$$

Hence, by subtraction, $1/2D = 29.060$, so that heritability is $29.060/40.350 = 72\%$.

Therefore, if the 5% earliest $F_2$ plants had been selected, the expected genetic advance from $F_2$ to $F_3$ can be calculated as:

$$G_s = (k)(\sigma_A)(\sigma_a^2 / \sigma_A^2) = (2.06)(\sqrt{40.350})(0.720) = 9.42 \text{ days.} \qquad (8\text{-}6)$$

Accordingly, the mean of the selected $F_3$ progenies is expected to be $21.2 - 9.4 = 11.8$ days. The actual advance in heading time in $F_3$ progenies derived from the earliest 5% of $F_2$ plants was 8.5 days. Thus, despite the numerous dubious assumptions made in formulating the prediction equations, the procedures followed in this experiment predicted the actual advance from $F_2$ to $F_3$ with some modicum of accuracy. The $F_2$ and $F_3$ data this suggest that most of the difference between $P_1$ and $P_2$ is due to a single gene pair, which in turn suggests that little more than the

TABLE 8-3 Dates of heading of the homozygous wheat varieties Ramona ($P_1$) and Baart 46 ($P_2$) and hybrid generations derived from the cross of $P_1 \times P_2$

| GENERATION | Days to Heading from an Arbitrary Date | | | | | | | | | | | | | $n$ | $\bar{x}$ | $s^2$ |
|---|---|---|---|---|---|---|---|---|---|---|---|---|---|---|---|---|
|  | 4–6 | 7–9 | 10–12 | 13–15 | 16–18 | 19–21 | 22–24 | 25–27 | 28–30 | 31–33 | 34–36 | 37–39 | 40–42 |  |  |  |
| $P_1$ | 4 | 21 | 60 | 48 | 20 | 4 | 2 |  |  |  |  |  |  | 159 | 12.99 | 11.036 |
| $B_1$ | 1 | 12 | 88 | 77 | 85 | 50 | 6 | 4 | 1 | 1 |  | 1 |  | 326 | 15.63 | 17.352 |
| $F_1$ |  | 1 | 2 | 20 | 83 | 51 | 12 | 2 |  |  |  |  |  | 171 | 18.45 | 5.237 |
| $F_2$ |  | 4 | 25 | 66 | 156 | 115 | 50 | 41 | 38 | 34 | 16 | 4 | 3 | 552 | 21.20 | 40.350 |
| $B_2$ |  |  | 4 | 34 | 49 | 47 | 45 | 61 | 41 | 26 | 6 | 1 |  | 314 | 23.28 | 34.288 |
| $P_2$ |  |  |  |  |  |  | 33 | 56 | 19 | 5 |  |  |  | 148 | 27.61 | 10.320 |

Data of 1953 season.

recovery of the parental genotypes may result from selection in later generations. It would be helpful to plant breeders if reasonably accurate predictions of genetic gain under selection could be made from such experiments. To test whether this is indeed the case for the apparently simple Ramona × Baart cross, several laborious follow-up experiments based on later, more informative generations were conducted.

The generations analyzed in the follow-up experiments fell into three groups: (1) additional early generations, (2) several later generations ($F_3$ to $F_7$) derived by repeatedly selfing randomly chosen $F_2$ and first backcross $F_2$ plants, and (3) several later generations ($F_3$ to $F_7$) in which selection for earliness (or lateness) was practiced in the $F_2$ and each succeeding generation.

The additional early generation materials included Ramona ($P_1$), Baart ($P_2$), the $F_1$ and $F_2$ generations, the backcrosses of the $F_1$ to $P_1$ ($B_1F_1$) and to $P_2$ ($B_2F_1$), as well as 25 $F_3$, 25 $B_1F_2$, and 25 $B_2F_2$ families obtained by selfing randomly chosen $F_2$, $B_1F_2$, or $B_2F_1$ plants. Large numbers (approximately 400–500 plants of each of these generations) were grown in field nurseries in each of two years; in these plantings each of the 25 $F_3$, $B_1F_2$, and $B2F_2$ families was represented by 17 individuals. Spacings were 30 cm between and within rows to ensure that only single plants within each family were measured. Distributions were nearly identical in all generations in both years; $P_1$ plants headed approximately 15 days earlier on average than $P_2$ plants, and the $F_1$ plants had virtually completed heading before $P_2$ plants began heading in all years. The heading times for $B_1F_1$ generation plants also differed little from those of either $P_1$ or $F_1$ in all years. Agreements between observed and theoretical distributions, as had also been the case in all previous early generation studies of this hybrid, were very close. This, again, suggests that the difference in heading date between $P_1$ and $P_2$ was due largely to a single major gene pair and, hence, that selection, especially for a heading date earlier than that of $P_1$, would be unlikely to be very rewarding.

The set of advanced-generation materials was derived from 200 randomly chosen $F_2$ individuals; nearly all of these randomly chosen $F_2$ plants headed within a day of the early or the late parent. Thereafter, a single selfed plant from each family of 17 plants was chosen, again by random methods, to perpetuate the family, and heading date was recorded for each plant in generations $F_4$ to $F_7$. The distributions of $F_4$, $F_5$, $F_6$, and $F_7$ family means for heading date differed from those of the earlier generations ($F_2$, $F_3$) in two main ways. The main difference was that many families that were more or less intermediate between the $P_1$ and $P_2$ appeared; also several families that were a day or more earlier than $P_1$ or later than $P_2$ appeared. Thus the distributions tended to be much less sharply bimodel than those of the $F_2$ generation. These observations were clearly at variance with the earlier conclusion that the differences in heading date between $P_1$ and $P_2$ were due almost entirely to only a single major gene pair that exhibited substantial dominance in the direction of earliness. Applications of multiple-range tests showed that at least 12 significantly nonoverlapping groups of family means were present among the 197 $F_7$ families. Assuming that nearly complete homozygosity had been achieved in generation $F_7$, these observations indicated that $P_1$ and $P_2$ probably differed by four gene pairs. With one gene pair, two genotypes, one carrying the major $P_1$ early allele and one with the major $P_2$ late allele, would be expected; thus, three or more additional gene pairs with smaller effects on heading date must be postulated to account for the appearance of recombinant genotypes intermediate between the parents, as well as those recombinant genotypes that that were, on average, about a day earlier or later than the parents.

The selected materials described earlier were derived from the 10% earliest and 10% latest among 420 $F_2$ plants. Thereafter, a single row (17 plants) of each of the 42 early and 42 late families was grown, and the earliest individual (in early families) and the latest individual (in late families) were saved in each family to perpetuate the family. Change in heading date under this type of selection was very rapid in early generations. Nearly all progeny of the late selections made in $F_2$ were later than $P_2$ in generation $F_3$, and in generation $F_4$ about half of the early families were earlier than the plants of generation $F_2$. The range from the earliest to the latest row for parent $P_1$ and for parent $P_2$ was about 20 days, whereas the range from the earliest to the latest selected families in $F_4$ was much larger, about 40 days. Additional changes occurred in the selected families in generations $F_5$, $F_6$, and $F_7$; in these generations individuals with intermediate heading dates disappeared completely and the distributions became bimodel, with one peak centered about one day earlier than $P_1$ and a second peak centered about 8 to 9 days later than $P_2$. Both of these two peaks were, however, substantially wider than those of $P_1$ or $P_2$. Clearly, selfing, combined with selection for earliness as well as for lateness, eliminated heterozygous genotypes that had phenotypes similar to or more or less intermediate between those of $P_1$ and $P_2$. At the same time, selection for earliness (or lateness) evidently also increased the frequency of genotypes that produced phenotypes earlier (or later) than $P_1$; this had the effect of broadening both shoulders of the early peak, whereas selection for lateness enhanced both shoulders of the late peak (skewness in the direction of lateness was, in fact, much more pronounced for the late peak). By later generations ($F_6$, $F_7$) the range in heading date from the earliest to the latest family had increased to nearly 45 days, primarily because of increases in the frequency of phenotypes later than those of generation $P_2$. Application of multiple-range tests to the combined random and selected data sets indicated the existence of 16 or 17 phenotypic classes. Assuming that each phenotypic class corresponded to only one homozygous genotype, at least 5 gene pairs with effects that vary from large (the major earliness-lateness gene separating Ramona and Baart) to 3 or 4 genes with much smaller effects must be postulated to explain heading date in generations $F_3$ to $F_8$ of the Ramona × Baart cross. Clearly, nearly all of the assumptions made in developing the original model were invalid, despite the nearly 3:1 distribution in generation $F_2$.

The genetic response equation $G_S = (k)(\sigma_A^2)(H)$ has often been used to predict genetic response. The equation sometimes predicts well for one generation; this was more or less the case from $F_2$ to $F_3$ in the Ramona × Baart study. Should the equation not predict well for even a single-generation transition, this may be due to inadequate estimates of any one or all of the three terms in the response equation. Thus, even though estimates of the variances of each of the $P_1$, $P_2$, and $F_1$ generations were consistent in the many different years in which the Ramona × Baart study was conducted, the variances of the three generations differed widely within each year. Hence, the estimate of environmental variance, namely, the mean of the variances of the three principal nonsegregating generations ($F_3$, $F_4$, $F_5$), is clearly suspect.

Moreover, selection was intense so that the estimates of values of $k$, which, of necessity, were determined each year from the very small tail of the curve in which estimates of variances were weak, are also suspect. This suggests that the moderately good prediction for gain from $F_2$ to $F_3$ in the earliest Ramona × Baart study was, at best, a fortuitous chance occurrence. The third term of the prediction equation, $\sigma_A^2$ is still more troublesome. The later-generation data ($F_3$–$F_7$) suggest that at least 4 and possibly more additional gene pairs were involved. In the case of five equally effective alleles at each of $N$ diploid loci, $3^N - 1 = 3^5 - 1 = 242$; thus a minimum of 242 parameters would necessarily have to be estimated to specify the genotypic values, provided that position effects, cytoplasmic effects, effects of genotype-environmental interactions, and other possible disturbances are ignored. Among the genetic parameters, 5 represent additive effects, 5 represent dominance effects, and the rest represent interactions among nonalleles. Thus, if 2, 3, 4, or 5 loci, rather than only 1 locus, are involved in the Ramona × Baart cross, $3^N - 1$ and 8, 26, 80 or, perhaps, even 242 genetic parameters, respectively, must be estimated to specify the effects of segregation in generations $F_3$, $F_4$, $F_5$. . . . Clearly, even such an apparently simple situation as that of heading date in the Ramona × Baart cross gives rise to immensely complex estimation problems. One way to circumvent the difficulties this complexity causes may be to self lines derived from random $F_2$ plants to near fixation by generation $F_6$ or $F_7$ to produce a random set of nearly completely homozygous inbred lines (Table 13-1). This is not to imply that a study of such materials would solve all of the complex problems engendered by segregation, especially in generations $F_2$, $F_3$, $F_4$, and $F_5$. A study of three materials might, however, give a better idea of the number of loci involved and the effects of such loci during segregating generations. As discussed earlier, this was clearly the case, but interpretations remained complex.

The experiment now to be considered was based on four isogenic lines of barley developed by C. A. Suneson of the U.S. Department of Agriculture (USDA), who crossed a long-selfed line (> 20 generations of selfing) derived from a single plant of an apparently highly homozygous population of Atlas barley, which has rough awns ($RR$) and white lemma ($oo$), with a second similarly long-selfed variety with orange lemmas ($OO$), and also with a third similarly long-selfed variety with smooth awns ($rr$). In the case of the orange lemma locus ($OO$) and the smooth awn locus ($rr$), 6 and 19 backcrosses, respectively, were made to Atlas during which $Oo$ (linkage group VI) and $Rr$ (linkage group VII) loci were held heterozygous. One or more generations of selfing followed each backcross in the two series, and intensive selection for the Atlas phenotype was practiced during the extensive program of backcrossing and selfing. After the final backcrosses, the homozygotes $ooRR$ and $OOrr$ were extracted during two generations of selfing and crossed to obtain the double heterozygote $RrOo$. The four possible homozygotes were then obtained by selection in appropriate $F_3$ families derived from a single plant of the double heterozygote $RrOo$. All crosses were made in such a way that Atlas cytoplasm was present in all four lines.

These four homozygous lines were expected to be either fully "isogenic," or very nearly so, for all loci excepting the two marker loci ($RrOo$) and very short closely linked segments on either side of the marker loci. The length of segment introduced in a series of $t$ backcrosses is expected to be $1/t$ crossover units with variance of $1/t^2$ crossover units on each side of the marker locus. Selfing reduces the length of the heterozygous segment at twice the rate of backcrossing because, under self-fertilization, effective crossing can occur in both the male and female parents.

If the selection practiced during the development of the isogenic lines is ignored, the half-lengths of chromosome segments associated with $o$ and $r$ can be estimated to be of the order of $2.9 \pm 2.9$ and $1.0 \pm 1.0$ crossover units, respectively. However, small phenotypic differences are readily discerned between and within such highly homozygous and phenotypically uniform families, and selection can be highly effective during a program of backcrossing and selfing of the type by which the lines of this experiment were developed. It is probable, therefore, that the map lengths of the segments associated with marker alleles $o$ and $r$ were smaller than indicated by computations in which it is difficult to take the effects of selection into account.

The following experiment was based on data obtained by testing stocks in which the two pairs of homologous segments occurred in all possible combinations. The critical question is whether the four marked segments were the same in each of the parental lines. If this was not the case, interaction effects might be confounded with the effects of segregation. In appraising this question, note that each of the four isogenic lines was descended from a single doubly heterozygous individual. If the two marked segments were indeed 6 and 2 crossover units long, as calculated, the probability that no crossovers had occurred in the $o$ and $r$ segments in the single parental plant of the four lines is about 0.94 and 0.98, respectively, and the probability of no crossovers in all four lines is approximately $(.94)^4(.98)^4 = 0.72$. Thus, even ignoring the intense selection that was practiced during the development of the isogenic lines, the probability that the integrity of corresponding segmental pairs was maintained during the final selection of the four parental plants (*OORR, OOrr, ooRR,* and *oorr*) ultimately tested considerably exceeds the probability that the segments were not identical.

The experimental materials on which measurements were taken were obtained by crossing the four isogenic lines (*OORR, ooRR, OOrr, oorr*) in all combinations, including selfs but excluding reciprocal crosses, to produce nine genotypes, *OORR, OORr, OOrr, OoRr, OoRr, Oorr, ooRR, ooRr,* and *oorr*. To obtain good stands and synchronous starts, the seeds obtained from these crosses were sown in flats and then transplanted into a field nursery at the second-leaf stage. The basic field design was a randomized complete block with 10 replications in which individual plants, spaced 45 cm apart, were the experimental unit. Six sets of the basic design (representing six replications) were sown on each of three planting dates about 1 month apart. Hence, each of the 9 genotypes was represented by 60 plants for each planting date (180 plants in total).

The significance of observed differences among genotypes was tested by standard analyses of variance for each of eight quantitative characters often considered to affect adaptedness and performance in barley: heading date, plant height, straw weight, number of spikes, weight of spikes, spike length, mean individual spike weight, and spike density. The variance ratios for genotypes in this experiment turned out to be highly significant (1% level) for six characters, significant (5 % level) for one character, and nonsignificant for one character (straw weight.); the straw weight character was not analyzed further. Variance ratios for planting dates were significant for all characters; however, the item genotypes × sets within planting dates was uniformly nonsignificant.

In analyzing the relative effects of genetic and environmental factors on the several traits, the observed phenotypic variances were partitioned into three components: (1) an environmental component, $\sigma_E^2$, reflecting plant-to-plant variability, including possible errors of measurement within replications, (2) a genetic component, $\sigma_G^2$ reflecting the individual effects of each of the 4 marked segments, and (3) a component, $\sigma_{G\times E}^2$ reflecting interactions of genotypes with within-set environments. Estimates of these parameters were then obtained by equating observed mean squares to mean square expectations and solving the relevant equations. These estimates revealed that the genotypic components of variance ($\sigma_G^2$) were always much larger than the components of variance due to genotype × environment interactions ($\sigma_E^2$). The mean squares for the components of genotypic variance were calculated from phenotypic values, and they are consequently inflated by components in $\sigma_E^2$ and $\sigma_{G\times E}^2$ that must be removed to obtain proper estimates of genotypic variances. Estimates of the components of genotypic variance, as expected, were much smaller for the isogenic materials than those calculated for the same characters in nonisogenized materials.

The next step in the analysis was to partition the genetic variance for each phenotypic class into its component parts. There are three possible genetic phases for the very short chromosome segment marked by locus *o* (*OO, Oo* and *oo*) and for the chromosome segment marked by locus r (*RR, Rr*, and *rr*). Considering the two loci (segments) simultaneously, nine combinations can be identified and estimated. The 8 degrees of freedom can be allocated as follows:

*A* = additive effect of locus (segment *o*)
*A* = additive effect of locus (segment *r*)
*D* = dominance effect of locus (segment *o*)
*D* = dominance effect of locus (segment *r*)
*A A* = interaction effect additive in o and additive in *r*
*A D* = interaction additive in o and dominance in *r*
*D A* = interaction dominance in o and additive in *r*
*D D* = interaction effect dominance in o and dominance in *r*

Tests of significance showed that additive effects (*A* and *A*) were significant in 11 of 14 cases (79%), that additive × additive interaction effects (*A A*) were significant in 6 of 7 cases (86%), followed by dominance × dominance effects (*D D*) in 4 of 7 cases (43%) and by additive × dominance effects (*A D* and *D A*) in only 2 of 14 cases (14%).

This experiment was based on two short chromosome segments that had been placed in an otherwise apparently highly isogenic background by a combination of extensive backcrossing and selfing. One of the segments (*o*) was probably smaller than 6 crossover units in length, and the other (*r*) was probably smaller than 2 crossover units. All of the 7 chromosomes of barley are more than 100 crossover units long. Hence, the o and r segments transferred into the isogenic lines probably represented < 1% of the total map length of barley, and, owing to the isogenicity of the genetic background, each of these two segments can reasonably be regarded as a miniature genetic system embedded within the ~99% of the Atlas genotype that had been rendered invariant by repeated backcrossing and selfing. Despite their small sizes, these two chromosome segments turned out to be genetically active for 7 of the 8 quantitative characters analyzed. Averaged over these 7 characters, the additive variance $(\sigma_A^2 + \sigma_A^2)$ accounted for more than half of the total genetic variance (65%). However, epistatic variance $(\sigma_{A \times A}^2, \sigma_{A \times D}^2, \sigma_{A \times D}^2, \sigma_{D \times D}^2)$ accounted for much of the rest (32%). This demonstration, that epistatic variance constituted a major part of the total variance for these two randomly chosen, very short segmental pairs, suggests that epistatic gene action may also be a feature of one and perhaps many of the very large numbers of comparable segmental pairs that would be expected to segregate in ordinary hybrids. Thus, genetic analyses of hybrids that segregate simultaneously for several, up to many, loci can be expected to be incredibly complex genetically, almost certainly much more complex than was the case in the present materials in which, by extraordinarily laborious efforts, the genetic situation had been reduced to two-locus (or very near two-locus) simplicity. It should also be noted that the two-locus analyses were probably deficient in several ways, largely because of economic practicalities. Some obvious deficiencies were that (1) phenotypic expressions of the four segments were measured in only one test year, with the result that year-to-year variation, which may be important, remained unknown; (2) only two loci (segments) were examined; hence, only main effects and first-order interactions could be examined and higher-order interactions could not be evaluated; and (3) although the experiment made clear that some characters (e.g., plant height) for the *o* segment were more active genetically than in the *r* segment, it provided little information concerning the effects of any of the four segments studied on adaptedness or on agricultural worth (i.e., on the questions that most concern plant breeders). As we will see later in Chapters 10 and 11, much more "cost-effective" methods of evaluating epistatic interactions among loci show that such interactions often play a highly important role in promoting adaptedness in nature and agricultural worth in cultivation. Clearly, however, attempts to estimate components of genetic variance following methods proposed by quantitative (biometrical) geneticists of the 1920s are not likely to be cost-efficient for plant breeders.

## PARENT-OFFSPRING RELATIONSHIPS

We now turn to ways of assessing responses to selection that have come to be widely used in practical breeding. Practicing breeders commonly inspect their populations carefully at various times throughout each growing season, judging individual populations, and often families within populations, or perhaps even single individuals within families for perceived overall worth. Many characteristics that contribute to overall worth are themselves genetically complex. Hence, the breeder's character "overall worth" can be expected to be, and, in fact, when carefully examined has nearly always been found to be, very complex genetically. Nevertheless, progeny of selected individuals are likely, owing to heredity, to resemble their parents in the next generation. If the agreement is very close, this suggests that genetic effects are relatively large and the effects of environment are relatively small. It also indicates that the breeder has estimated, by "eyeball" methods, truly important components of the genetic advance equation with some precision. At the other extreme, failure of progeny to resemble their parents suggests that environmental effects are relatively large, as compared with hereditary effects, or that breeders' eyeball assessments of the genetic components of "overall worth" were probably faulty. Note that this approach to parent-offspring relationships is essentially empirical: it assesses parent-offspring relationships directly and makes no genetic assumptions, except that it does involve the concept of separating environmental effects from genetic effects. Sometimes breeders make numerical measurements on parents and on offspring, and calculate regressions from their measurements. Statistically, such regressions avoid some of the numerous problems associated with the partition of variances, but such regressions are not entirely free of problems. Although parent-offspring regressions have been used more widely by animal breeders than by plant breeders, many experienced plant breeders are convinced that eyeball assessments of overall worth are more reliable than numbers that, unfortunately, must usually be obtained by very laborious measurements. In the next chapter we examine the conversion of the wild ancestor of corn by prehistoric farmers who used eyeball selection to convert a wild Central American grass of questionable value as a cereal that has subsequently become one of the most important cereal crops of the world. Note, however, that this process required not only much effort but also centuries to accomplish.

# NINE

# *Evolution During Cultivation*

The evidence is compelling that many, and possibly all, modern cultivars are now better adapted and more productive in the agricultural habitats in which they are presently grown than when they were when they were first domesticated. Although *adaptedness in different habitats in nature is complexly inherited, adaptedness in cultivation is almost certainly still more complexly inherited.* In addition to carrying out essential biological processes efficiently, modern cultivars must satisfy numerous production requirements of farmers as well as the numerous quality preferences of the marketplace. These attributes of production and quality sometimes run counter to underlying biological processes, thus complicating the tasks of plant breeders. Despite such complications, it is usually obvious from even cursory comparisons of modern cultivars with their wild ancestors *that a great deal of genetic progress has been made in improving the performance of crops under modern agricultural conditions,* especially in productivity (yield and product quality), during the hundreds or even thousands of years that many crop plants have been under improvement in agriculturally diverse conditions in many different areas of the world. There can be no doubt that landraces of major crops perform their agricultural tasks much better than their wild ancestors and that modern cultivars are nearly always conspicuously more successful, at least in modern high-input agricultures, than the landraces they supplanted. Such observations *notwithstanding, it is now commonly acknowledged that numerous factors* (genetic, environmental, agricultural, and economic) affect suitability in cultivation and that the effects of these factors are difficult to evaluate, not only individually but also collectively. Accordingly, our understanding of what actually happened genetically during domestication itself and during the hundreds or even thousands of years that crop plants have been undergoing improvement under agricultural conditions in many different places in the world, is far from complete. Nevertheless, it is clear that preagricultural societies that depended largely on wild plant foods *emerged in several different areas during the Paleolithic Age and that in each such* area dependency was dominated by relatively few species. Once an archaic society had become committed to a suite of wild plants as dietary staples, the stage was set for the next step—domestication. Much of the evolution of the human species took place during the Ice Age, the Pleistocene, but the emergence and development of agriculture-based societies took place in the Holocene, the post–Ice Age world, and it is the 11,000 years or so of that period that are of the most interest agriculturally. Cowan and Watson (1992) infer two main themes in agricultural origins. One they refer to as pristine, involving the gradual addition of crop plants descended from the native flora of an area, to an existing foraging-based economy. They regard the Near East and southern Mexico as two of the best examples of this process. There are, in contrast, two sorts of situation in which agricultural economies developed secondarily. In one such situation a suite of domesticated plants that was developed elsewhere was gradually introduced into an area previously dominated by foraging; whether crop by crop, or en masse, agriculture slowly replaced foraging in the recipient area. Cowan and Watson regard Europe and the eastern American wood-lands as prime examples of this secondary acquisition of agriculture by importation from elsewhere of previously domesticated plants. Secondary introductions some-times also featured the introduction of exotic crop plants into human cultural systems that had already become, at least in part, dependent on agricultural plants. Once such a secondary introduction occurred, the introduced domesticated plant sometimes replaced the old suite of crop plants to become the dominant compo-nent of the future agricultural economy of the region. Such replacements occurred in eastern North America, where the seed plants of eastern woodlands' pristine complex were slowly replaced by corn, and in Japan, where a multicrop agriculture came to be dominated by rice within a few hundred years after this crop had been introduced to southwestern Japan from mainland Asia. Although the early history of the domestication of corn is debated and its spread into South America is open to argument, its spread northward in Mexico and into North America is fairly well known. Corn was introduced into areas east of the Mississippi River about 2000 B.P. at a time when the American Agricultural Complex of the eastern woodlands, based on several endemic starchy and/or oily small-seeded annuals (sunflower, *Helianthus annuus;* little barley, *Hordeum pusillum;* goosefoot, *Chenopodium berlandieri*) as well as nuts from several perennial trees, was already well developed. For the next six or seven centuries corn was grown by eastern woodlands Indians, but at a level hardly detectable in the paleoethnological record. Beginning about 1000 years B.P. corn production expanded explosively, and corn rapidly largely replaced the old endemic cropping system in the eastern woodlands. The more or less 1,000-year gap between the time corn was first introduced into the region and the time it became the mainstay of agricultural economies can be explained in ecogenetic terms. Corn is a subtropical domesticate from Mexico, a domesticate adapted to growing in climatic conditions very different from those of eastern North America. Archaeological remains of early corn in eastern North America are not numerous, but they depict a diverse gene pool that various human societies in different parts of the region manipulated by selection for desired characteristics. By about 1000 years B.P. two major types of corn were being grown east of the Mississippi River, one adapted to the warm Southeast and the other to the cooler mean temperatures and the shorter day lengths of Canada and central North America. Not until these two major types had evolved could corn become an effective competitor to the eastern complex of annuals. However, once many local types of corn were in place, corn rapidly gained ascendancy in the indigenous crop complex in the northern and central regions of North America. By the time of Columbus (~500 B.P.) about

200 or more distinctive varieties of corn had been developed by Indians in North America.

Although there is little consensus on the origin and the earliest evolution of corn in cultivation, there is reasonable agreement that corn was first domesticated about 10,000 years B.P. somewhere in tropical or subtropical Mesoamerica, probably in present-day southern or southwestern Mexico, at latitudes of about 15°N. This is an area of great environmental and ecological diversity where sharp variations in altitude tend to offset latitude in sharply differentiated adjacent areas in which agricultural and foraging societies existed side by side; thus, there have been difficulties in determining precisely where or when domestication of corn began. Several features have been cited, suggesting that the cultivated type is descended from wild teosinte *Zea mexicana* (syn. *Euchleana mexicana*): (1) the frequent hybridizations that occur between corn and teosinte under natural conditions, (2) the identical chromosome number (10 pairs) and the similar chromosome structure of teosinte and corn, and (3) numerous anatomical and morphological similarities. It is also not clear whether corn was originally domesticated as a cereal (because the seeds of wild teosinte are small and very hard, they probably were not appealing as a cereal), as a vegetable crop (corn tassels are edible as a vegetable), or as a ceremonial crop (corn had religious as well as ceremonial significance in many early societies). It is therefore possible that these diverse early usages may have muddled our understanding of the effects of the differing sequences of selection that possibly occurred in the early evolution of corn during its domestication and as the species migrated southward into South America and northward into North America. Corn appeared in the desert borderlands of northern Mexico about 1000 B.P. After its arrival there it spread rapidly northward, where it became a primary source of carbohydrates for the next 3,000 years or more, suggesting that selection emphasized its use as a cereal during that period. Corn cultigens then spread eastward and northward in North America where, as noted earlier, it remained a minor cultigen during the six centuries from about 1800 B.P. to about 1200 B.P. During this six-century period efficient food-producing economies based primarily on several North American species became increasingly important in North America (Smith, in Cowan and Watson 1992). During the next three centuries (from about A.D. 800 to A.D. 1100) this trend took a new direction in which the use of corn expanded explosively, making it by far the dominant component in northeastern American agriculture about 900 B.P. Thus, increasingly corn-dominated economies encompassed most of the formerly largely foraging societies of the coastal plains of the Gulf of Mexico and the Atlantic coast of North America, as well as the midlatitude and northerly farming zones of the midcontinent. According to Smith (1993) the shift to corn-centered agriculture from about 300 to 1100 B.P. in much of northern Mexico and North America was not merely a matter of developing improved types of corn. The shift was embedded within a broader and uniquely North American process of a major social transformation that emphasized agriculture. By the time of Columbus (~ 500 B.P) at least 200 varieties of flint corn had been developed by American Indians. Flint corns have small, hard, rounded seeds in which a small amount of soft endosperm is enclosed within a vitreous outer layer. Columbus took flint corns to Europe (from Cuba) about A.D. 1500, and by A.D. 1600 flint corns had spread across Europe, Africa, and Asia.

Archaeological evidence indicates that dent corns had appeared in north-eastern Mexico by 500 B.P. and that they appeared soon thereafter in coastal southeastern North America. Dent corns have a substantial amount of soft starch in their kernels enclosed by a thin outer layer of hard starch, and their kernels become indented upon drying at maturity. Dent corns are usually considerably higher yielding than flint corns and superior in several other ways, but they are not as well adapted to short seasons as flint corns. Flint and dent corns evidently stayed separate for some time in southeastern North America, apparently because they were distinctly different in so many traits. However, by about A.D. 1800 North American colonists found that hybrids between the two types were advantageous; flints provided genes for earlier maturity, and dents for higher yield (Doebley et al. 1988). These findings coincided with rapid westward and northward migrations in the United States into areas requiring corn types with earlier maturities and greater resistance to stresses (e.g., lower rainfall and higher summer temperatures) than dent corn. By the late 1830s, Tennessee had become the leading corn-growing state, soon followed by Kentucky and Virginia. By the late 1870s, Illinois and Iowa led in corn production. A variety adapted to Tennessee that had been moved to Ohio in the 1830s had, by the late 1870s, about 50 generations of natural selection for conditions of longer days, shorter seasons, somewhat less rainfall, and, in many cases, higher yielding conditions than in Tennessee. Human selection was no doubt also important: it was practiced for earlier flowering and for earlier ripening combined with high yield, good ears, and many other characteristics desired in various places in the multistate areas in which such characteristics were advantageous. It has been estimated that more than 500 open-pollinated varieties of corn existed in the United States and Canada in 1840, and more than 1,000 open-pollinated varieties by 1920, nearly all of which were ultimately selfed to develop inbred lines in the 1930s or 1940s. Many breeders of inbred lines of the 1920s wanted the noticeable local adaptedness that had been important for open-pollinated varieties. However, it turned out that inbred lines derived from widely adapted open-pollinated varieties ultimately prevailed over inbred lines derived from open-pollinated varieties that were conspicuously well adapted only locally. How were widely adapted open-pollinated varieties of corn developed? The development of Reid Yellow Dent is the leading example.

Reid Yellow Dent, developed by James Reid, became the most popular open-pollinated variety in the United States. James Reid's father, Robert Reid, moved his family from Ohio to Illinois late in the spring of 1846. He brought along a dent corn, which he planted late that year and from which he harvested a crop that included many immature kernels. In 1847 the seeds harvested the preceeding fall produced a stand that was judged to be poor but too good to be abandoned. Robert Reid replanted the missing hills with an early native Indian flint corn. His replanting allowed the two types of corn with different maturities to intercross, and he used this cross-pollinated seed for planting the next crop. In following generations he selected for maturity appropriate to north-central Illinois (one parent of his highly mixed population was too early and the other too late for the local conditions). He also taught his son, James, his philosophy of corn selection: emphasis was on medium-size ears, medium maturity, bright yellow kernels with smooth grain, and 18 to 22 kernel rows well filled over the ends of the cob, giving a high shelling percentage. He selected mature dry seed in the field at harvesttime. It was not until James Reid had practiced selection for more than 40 additional years that he won a blue ribbon for the best corn at the Illinois State Fair in 1891 and a gold medal at the World's Fair in 1893. The seed business James Reid started was modest because strong selection continued: only 5% of the ears from his highest-yielding field were harvested, providing only about 10,000 kilos of seed per year. Selection continued to emphasize proper maturity as well as high yield. Reid Yellow Dent became the most popular corn variety across nearly all of the United States. The variety must clearly have received an imposing array of adaptedness genes from its very differently adapted flint and dent parents to have become the most popular open-pollinated variety over such a vast and highly diverse area. With the advent of hybrid corn in the late 1920s and 1930s, outstanding inbred lines were selected directly out of open-pollinated Reid Yellow Dent. Many widely grown open-pollinated varieties from which superior inbred lines were subsequently selected can be traced back to Reid Yellow Dent. In addition, Stiff Stalk Synthetic (SSS), developed in the early 1930s from intermatings among 16 inbreds, most with Reid Yellow Dent background, contributed a number of inbreds that were widely used in the United States, as well as in several popular early Canadian and European inbred lines. In addition, another popular widely grown open-pollinated variety, Lancaster Sure Crop, also derived from flint-dent crosses, has provided many useful inbred lines on selfing, including many lines that "nick" well with inbreds from Reid Yellow Dent.

Although most of the early breeders of inbred lines attempted to develop such lines from locally adapted open-pollinated varieties, this ultimately turned out not to be the thing to do. The record shows that a better course to follow was to develop inbred lines from open-pollinated varieties that were not only highly successful locally but also widely adapted. Inbreds developed from such sources much more frequently triumphed worldwide than inbreds from highly successful but narrowly adapted open-pollinated varieties. This indicates that there are genes, or multilocus combinations of genes, that lead to wide adaptedness.

This chapter has given a broad view of the evolution of corn under cultivation with a minimum of specifics. In the next two chapters (Chapters 10 and 11) we will explore in detail genetic aspects of the development of adaptedness in nature and in cultivation, and in Part III we will be concerned, in large part, with specific breeding plans designed to produce maximally productive varieties with both appropriate specific and broad regional adaptations.

# TEN

# Marker-Assisted Analysis of Adaptedness
# in Nature

Adaptedness in nature, as well as in cultivation, is complexly inherited and much affected by environment. The genetic mechanisms that have led to improved adaptedness and to enhancement in Darwinian fitness have, consequently, for reasons discussed in Chapters 6 to 9, been difficult to identify, and their effects have been difficult to measure precisely. Accordingly, until very recently, we have had little reliable information concerning the details of the genetic mechanisms responsible for the improvements in adaptedness that have occurred in nature, during domestication, during evolutionary diversification under cultivation, or as a result of modern breeding practices. In recent years it has been discovered that changes in the frequencies of discretely distinguishable alleles and multilocus genotypes of electrophoretically detectable variants, in natural as well as in cultivated populations, allow more precise tracking of adaptedness alleles and genotypes over generations. This, in turn, allows genetic analyses of the evolution of adaptedness and other complexly inherited quantitative traits to be interpreted in terms of selection acting on only one or, often, jointly on several to many different chromosome segments, each marked by individually distinguishable alleles of specific genetic loci. The outcome of such selection differs from habitat to habitat, leading to clearly recognizable patterns of multilocus variability in different environments. Thus, marker-assisted analyses of complexly inherited traits have become feasible, and such analyses have in recent years increasingly replaced traditional biometrical procedures such as those described in Chapter 8 in regard to applied plant breeding. The literature in marker-assisted genetic analyses is now too extensive and technical to treat comprehensively in a general plant-breeding textbook. Consequently, discussion of marker-assisted analyses in this chapter will focus first on the genetic changes that have occurred in two extensively studied wild species—diploid *Avena hirtula* (*Ah*, $n = 7$ chromosomes) and its autotetraploid descendant, *A. barbata* (*Ab*, $n = 14$ chromosomes)—as these wild species evolved in nature over thousands of generations in the Mediterranean Basin and the Middle East, as well as in recently established wild colonial populations of *Ab* in North America. In the next chapter we will consider evolutionary changes that have occurred in such marker loci during the domestication and breeding of three representative cultivated species: (1)

diploid cultivated barley, *Hordeum vulgare*, ssp. *vulgare* ($n = 7$ chromosomes), descended from wild *H. vulgare*, ssp. *spontaneum* ($n = 7$ chromosomes); (2) diploid cultivated corn (maize), *Zea mays* ($n = 10$ chromosomes), descended from wild teosinte, *Zea mays*, ssp. *mexicana* ($n = 10$ chromosomes); and (3) hexaploid cultivated oats, *Avena sativa* ($n = 21$ chromosomes), descended in turn from diploid wild *A. hirtula* ($n = 7$ chromosomes) and tetraploid wild *A. barbata* ($n = 14$ chromosomes) via wild hexaploid *A. fatua* ($n = 21$ chromosomes). Cultivated hexaploid oats and cultivated barley, in common with their wild ancestors, are heavily self-pollinated ($\sim 99\%$) and thus serve as representatives of inbreeders, whereas corn, also to be considered in Chapter 11, is largely outcrossing ($\sim 90\%$ outcrossing vs. 10% selfing) and thus serves as a representative of frequently outcrossing species. It should be noted in particular that modern corn breeding features heavy inbreeding; hence, this change in mating system illustrates the increasing possibilities that were presented to corn breeders when they adopted inbreeding as a major component of modern corn-breeding methods.

## *A. hirtula (Ah) and A. Barbata (Ab)*

Diploid *Ah* ($N = 7$ pairs of chromosomes) is indigenous to the Mediterranean Basin, where it usually occurs in sparse and more or less disjunct stands. *Ah* is nearly always less robust than wild *Ab*, its autotetraploid descendant ($N = 14$ chromosomes), but the two taxa are sufficiently similar phenotypically that, until recently, they could be distinguished positively only by chromosome counts. It has now been found they can also be distinguished unambiguously by means of electrophoretic profiles: *Ab* is conspicuously more variable both electrophoretically and genetically than *Ah*. *Ab*, which regularly forms 14 pairs of chromosomes, is a completely diploidized autotetraploid derived presumably from *Ah* (Hutchinson et al. 1983). *Ab* is much more widely distributed than *Ah*. Cultivated *A. sativa* ($n = 21$ chromosomes) behaves as a completely diploidized hexaploid derivative of *A. hirtula* and *A. barbata*; it is very widely adapted and is grown in cultivation in many parts of the world. *Ab* occurs in massive stands throughout the Mediterranean Basin across the Middle East to Nepal, and it has also been a highly successful colonizer worldwide in areas with Mediterranean-like climates. Autotetraploid *Ab* is clearly much more widely adapted and robust than *Ah*.

Wild *Ab* was accidentally introduced to California from southwestern Spain as a weed perhaps $\sim 150$ generations ago, but *Ah* has not been found in California. Most plants of *Ab*, which is very widely distributed in California, occur in two major climatic zones, a semiarid zone (250–350 mm annual rainfall) that includes nearly all of southern and central California, and a large generally mesic zone (350–1,500 mm annual rainfall), which extends about 1,000 km northward in the Pacific coastal ranges of California from $\sim 36°N$ to mesic northern Oregon ($\sim 45°N$). It was discovered in the 1930s that nearly all populations of *A. barbata* in the xeric zones of California are monomorphic for two variants (dark lemmas and glabrous leaf sheaths), both later found to be monogenetically inherited, whereas nearly all populations in mesic zones are monomorphic for light lemmas and hairy leaf sheaths, each also monogenically inherited. However, mesic patches located within generally xeric zone sites (generally located within cooler, higher-elevation, higher-rainfall sites), as well as xeric patches generally located within mesic zone sites (often on locally xeric steep southwest-facing slopes) were often polymorphic for these two pairs of variants. These contrasting ecogeographical patterns suggested that the two morphic pairs might "mark" and perhaps also contribute to the adaptedness of two "ecotypes" (*sensu* Turesson 1922) that presumably developed in nature in response to contrasting selection pressures imposed by the xeric versus mesic habitats. Experiments to test this hypothesis were carried out during the 1930s in several internally heterogeneous sites, each ~1 hectare in area, located about 100 km southeast of San Francisco in an area in which the xeric and mesic zones form numerous interfaces with each other. Small subsites within each internally hetero-geneous site were classified into one of five categories, extending from xeric (Category 1) through intermediate categories, to mesic (Category 5). Assignment to these five categories was based solely on various physical features of each subsite, such as slope, orientation to the sun, and edaphic characteristics. Each year during the 1930s, approximately 100 plants from many such subsites were classified for the simply inherited lemma color and leaf sheath variants and measured for several quantitative traits such as height, aboveground dry weight, and numbers of seeds produced on the primary tiller. The critical finding was that typical xeric and mesic subsites were always monomorphic, or very nearly so, for the single-locus dark-glabrous or the light-hairy morphs, respectively, but that intermediate-xeric, inter-mediate, or intermediate-mesic subsites were polymorphic for the two simply inherited markers in frequencies, correlated with degree of xerism. Another critical finding was that several quantitative traits measured were variable within all sub-sites and that the frequencies of the quantitative traits were significantly correlated statistically with degree of xerism. These results suggested that addi-tional simply inherited Mendelizing markers were likely to be more informative in future studies of natural ecotypic differentiation than continuously varying, difficult-to-quantify metrical characters. When ecogenetic studies were re-sumed in California following World War II they were largely confined to crop plants, especially barley and Phaseolus beans. This was because many highly isogenic simply inherited markers, derived by backcrossing or recombinant inbreeding methods, were available in these species. The most useful break-through respecting simply inherited markers came with the introduction of enzyme electrophoresis to plant genetics (Schwartz 1960). In short order large numbers of technologically advantageous electrophoretically detectable co-dominant allozyme markers had been identified and pressed into service in ecogenetic studies of several crop plants and their wild ancestors, including *Ab*, *Ah*, *wild barley* (*Hordeum vulgare* ssp. *spontaneum*), and corn (*Zea mays*). How-ever the studies carried out in the 1960s, 1970s, and 1980s usually dealt with only one or two of the several genetic mechanisms, usually mating-system mechanisms, that were ultimately found to be jointly responsible in shaping the internal genetic structure as well as the overall adaptive landscapes of the various species studied. Consequently, in the discussion to follow, attention will be focused on the results of a few recent studies that have taken into account simultaneously the genetic mechanisms that had been identified in earlier studies as most important in evolutionary change.

Table 10-1 gives a sample of allelic and genotypic frequencies for 14 representative codominant allozyme loci in 10 Spanish diploid *Ah* and 50 Spanish tetraploid *Ab* populations, distributed as shown in Figure 10-1. Cytogenetic analyses have established that *Ah* forms 7 bivalents whereas *Ab* forms 14 bivalents with great regularity; this supports earlier evidence that *Ab* is a fully diploidized tetraploid derived from *Ah*. Genetic studies (Hutchinson et al. 1983) have established that *Ab* does, in fact, behave genetically as a fully diploidized tetraploid—that pairing is fully preferential within each of the two sets of 7 chromosomes ($2n = 14$ chromosomes in *Ab*) and that exchanges of genetic materials do not occur between corresponding chromosomes of the two homologous sets of 7 chromosomes. Recently it has been established that meiosis is chaotic in colchicine-produced autotetraploids of *Ah*; many trivalent and quadrivalent associations are observed at meiosis in such newly formed tetraploids—so that separation of daughter chromatics is irregular in anaphase and fertility is very low (effectively zero). It is difficult to imagine that any of the newly formed and highly infertile autotetraploids could have produced sufficient numbers of progeny to become established within

TABLE 10-1  Single-locus duplex and quadriplex frequencies in 10 Spanish diploid (*Ah*) populations and 50 Spanish tetraplid (*Ab*) distributed as shown in Figure 10-1 (adapted from Allard et al. 1993)

| | DIPLOID | | | TETRAPLOID | | |
|---|---|---|---|---|---|---|
| LOCUS | DUPLEX | f | NO. SITES IN WHICH FIXED | QUADRIPLEX | f | NO. SITES IN WHICH FIXED |
| *Pgd2*[a] | 11 | 1.00 | 10/10 | 11,11 | 0.12 | 1/50 |
| | | | | 11,22 | 0.84 | 24/50 |
| | | | | 22,33 | 0.04 | 0/50 |
| | | | | | 1.00 | |
| *Mdh1* | 11 | 1.00 | 10/10 | 11,22 | 1.00 | 50/50 |
| *Est1*[b] | 11 | 0.07 | 0/10 | 33,33 | 0.15 | 1/50 |
| | 22 | 0.24 | 0/10 | 55,55 | 0.14 | 0/50 |
| | 55 | 0.44 | 2/10 | 11,33 | 0.04 | 0/50 |
| | 66 | 0.03 | 0/10 | 22,33 | 0.19 | 2/50 |
| | 77 | 0.19 | 0/10 | 22,55 | 0.15 | 1/50 |
| | | 0.97 | | 33,55 | 0.04 | 0/50 |
| | | | | 33,66 | 0.03 | 0/50 |
| | | | | 55,77 | 0.20 | 4/50 |
| | | | | | 0.94 | |
| *Pgm1*[c] | 11 | 1.00 | 10/10 | 11,11 | 1.00 | 50/50 |

[a] Loci *Pgd2*, *Mdh3*, *Acp1*, *Acp2*, and *Pgd1* have similar patterns.
[b] Loci *Est1*, *Lap1*, *Prxl*, and *Pgi* have similar patterns.
[c] Loci *Pgm1*, *Got2*, *Mdh2*, and *Got1* have identical patterns.
[f] Estimates based on 754 diploid plants (1,508 alleles) and 4,751 tetraploid plants (19,004 alleles).

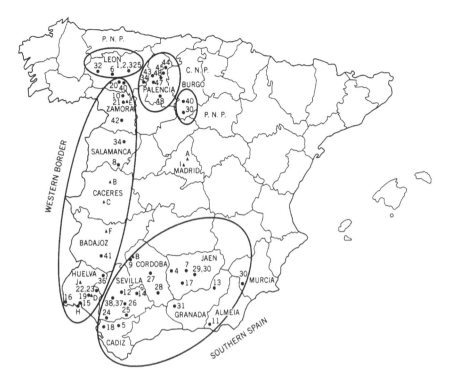

FIGURE 10-1   Ten Spanish diploid *Ah* and 50 Spanish tetraploid *Ab* populations.

the natural populations of *Ah* in which the tetraploids had apparently arisen. This suggests that some mechanism for suppression of pairing between homologues of the two genomes must have developed soon after at least one of the many events of polyploidization that must have occurred in the Mediterranean Basin over the centuries. However, sooner or later at least one such event must have occurred such that (1) this event quickly led to fertile diploidized tetraploid plants, whose progeny became established in the local ecosystem occupied by *Ah* and (2) these now fertile autotetraploid plants not only largely replaced diploid *Ah* locally but also spread, greatly extending the range of the diploidized tetraploid far beyond that of its diploid ancestor. Ladizinsky (1973) presented evidence that suppression of pairing in *Ab* is indeed due to a simple genetic mechanism that prevents pairing between any of the 7 homologs in the 2 genomes, thus in one step converting the infertile autotetraploid into a fertile, fully diploidized tetraploid that regularly forms 14 bivalents. However, $F_1$ hybrids occur rarely as a result of natural crosses between diploid *Ah* and tetraploid *Ab*, and when such hybrids occur, they are completely or nearly completely sterile. Meiosis is chaotic in artificially produced $F_1$ hybrids between *Ah* and *Ab*, and such hybrids are also nearly completely sterile. It therefore seems likely that *Ab* and *Ah* have been effectively isolated reproductively and that they have existed as effectively completely independent species subsequent to the production of *Ab* by polyploidization from *Ah*, presumably many thousands of generations ago.

## Allelic and Genotypic Frequencies in Ah and Ab

Table 10-1 gives single-locus duplex and quadriplex frequencies for 14 representative codominant allozyme loci in 10 present-day Spanish diploid (*Ah*) and 50 Spanish tetraploid (*Ab*) populations, distributed geographically as shown in Figure 10-1. Considering the diploid (*Ah*) populations first, the genotype of all 754 Spanish plants examined was found to be uniformly homozygous for allele *1* (i.e., genotypically *11*) for three loci (*Pgm1, Got2, Pgd2*) and for allele 2 (i.e., genotypically 22) for one locus (*Mdh1*); these alleles were named in the order in which they were discovered in the tetraploid (*Ab*). Each of these four loci was thus completely monomorphic (F = 1) in all of the 10 Spanish diploid populations examined. Locus *Mdh2* differed only slightly. Among the 754 Spanish diploid plants examined from 10 sites, 751 plants were genotypically 11 and 3 plants (all from a single site) were genotypically *22*: thus, 9 of the 10 different sites sampled were monomorphic for duplex *11*, whereas 1 site was weakly polymorphic for duplexes *11* and *22*. Allelic frequencies for locus *Mdh2* were F = 0.996 11: F = 0.004 *22*; these five loci (*Pgm1, Got2, Pgd2, Mdh1*, and *Mdh2*) have also been found to be completely monomorphic, or very nearly so, in all other diploid populations that have been sampled in the Mediterranean Basin. The predominant alleles of each of these five loci evidently encode for some essential function such that they confer survival ability superior to that of any and all other alleles of these five loci that have arisen by mutation during the entire evolutionary history of *Ah*. Evidently, none of the other alleles that have arisen in the diploid have been competitive with the presently predominant alleles, and not one among such mutant alleles of these loci have survived except in very low frequencies (e.g., allele 2 of *Mdh2*, discussed in this paragraph). Present allelic frequencies thus appear to provide a biologically meaningful measure of long-term survival values. Accordingly, the long-term survival values of the predominant alleles of loci *Pgm1, Got2, Pgd2, Mdh1*, and *Mdh2* appear to be effectively F = 1.00, and the long-term survival values of all other alleles that have arisen at these loci through mutation appear to be very close to F = 0 in each of the 10 Spanish *Ah* populations sampled, as well as throughout the entire range of distribution of the diploid throughout the Mediterranean Basin. The selective values (Chapter 8) of these predominant alleles are not necessarily F = 1.00 in all generations, but almost certainly they have been much higher, on average, than those of other alleles that ultimately were eliminated or reduced to very low frequency. Similarly, the selective values of low-frequency alleles (perhaps including those of some nonsurvivors) were not necessarily close to zero, but they were almost certainly lower, on average, than those of the ultimately predominant alleles.

The pattern of allelic variability differed somewhat for locus *Got1*. Although allele *1* of this locus was present in high frequency (F = 0.951) overall in *Ah* in Spain, a second electrophoretically delectable allele (allele 2) was found in one of the 10 Spanish sites assayed and allele *2* was, in fact, more frequent at that site (F = 0.712) than allele *1* (F = 0.288). Thus, present frequencies suggest that long-term overall survival values of alleles *1* and *2* of *Got1* are effectively ~0.95 and ~0.05, respectively, in the diploid in the various environments of Spain and that the long-term survival values of all other alleles that have arisen by mutation at these loci over the centuries are zero or close to zero. Allele 2 and a third allele (allele 3) of *Got1* have been found in occasional populations throughout the range of distribution of *Ah*, always in polymorphic association with allele *1*. Thus, alleles *2* and *3* of locus *Got1* apparently are not ill-adapted transients on their way to complete elimination by selection; they almost certainly make a positive, but usually minor, contribution to overall population adaptedness in a few habitats.

Patterns of within-site variability in *Ah* in Spain and elsewhere in the Mediterranean Basin and the Middle East indicate that allelic diversities are generally somewhat higher for loci *Mdh3*, *Acp2*, *Pgd1*, and *Acp1* than is the case for *Got1*. In Spain (Table 10-1) one allele of each of these four loci was present in high frequency $(0.6 < F < 0.90)$, and this most frequent allele was either predominant or fixed in most sites. However, some sites were polymorphic for a second, and sometimes for a third or even occasionally for a fourth, allele; moreover, one or another of these additional alleles was sometimes predominant or even fixed in occasional sites. This pattern of allelic variability suggests that one of the alleles of these five loci is nearly always superior to all others, but that there are occasional environments in which the long-term survival values of generally less frequent alleles are superior to those of the usually predominant allele. The presence in modest frequency of 2 (sometimes 3 alleles) of a single locus within some sites suggests that polymorphism may, in fact, improve overall population fitness in such sites.

The four remaining loci, *Pgi1*, *Prx1*, *Lap1*, and *Est1*, were found to be extensively polymorphic (4 to 7 electrophoretically distinct alleles) in the diploid in Spain (Table 10-1) and at most other sites in the Mediterranean Basin and the Middle East. Three or more alleles of each of these four loci were present in at least intermediate frequencies in about half of the 10 Spanish sites, as well as elsewhere throughout the range of the diploid. Clearly, the population genotypes that lead to optimum fitness in nearly all sites feature mixtures of at least 2 (usually more) alleles of these 4 loci.

Turning to the tetraploid, note that allelic diversity is much greater in *Ab* than in *Ah*; for example, *Ab* has 3 alleles of *Pgd1*, but only 1 allele of this locus has been found in *Ah* (Table 10-1). Counts made on all 14 loci show this is uniformly the case in *Ab*: 52 alleles (3.7/locus) were observed in *Ab* but only 38 (2.7/locus) in *Ah*. However, only 29 of the 38 alleles of *Ah* were present in *Ab*, indicating that 9 alleles of *Ah* were lost during or following the polyploidization process. This raises two important questions. The first is, which alleles were lost and which alleles survived? The answer to this question is clear-cut in the diploid versus the tetraploid *Avenas*. The 9 least-frequent alleles of *Ah*, which ranged in frequency from 0.001 to 0.070, were all lost, whereas the 29 more frequent alleles, which ranged in frequency from $F = 0.11$ to 1.0, all survived (loss of rare alleles is common during and/or following various momentous evolutionary events, such as during the formation of polyploids during migration to different habitats, or during their being taken into cultivation). The second question is, where did the 23 alleles $(52 - 29 = 23)$ of the tetraploid not present in the diploid come from? Two facts—first, that *Ab* behaves as a fully diploidized tetraploid, and second, that all of the "new" alleles of *Ab* were found in nonsegregating heteroallelic quadriplexes—suggest the answer. Locus *Pgd2* (Table 10-1) serves as a model for the sequence of events that appears to account for the greater allelic diversity of the tetraploid. Diploid *Ah* apparently is monomorphic (Table 10-1) for allele *1* of *Pgd2* in Spain (also worldwide); consequently, the original quadriplex formed following all polyploidization events was almost certainly *11,11*. Theory (Wright 1937) indicates that the probability is very small that any mutants among the many that come into existence over long periods of time will become established in small diploid populations, especially in small diploid populations that reproduce by self-pollination. Yet the likelihood of establishment is much higher in a tetraploid population for any one of three reasons: First, mutations are equally likely in either of the two genomes; second, population sizes are typically larger in the tetraploid than in the diploid. Hence, owing to these two reasons, more than twice as many mutations of allele *1* to allele *2* are likely to occur in *11,11* quadriplexes of the tetraploid than in the 11 duplexes of the diploid. The third reason is, however, almost certainly much more important: In a diploidized tetraploid such as *Ab*, each mutation will lead to a *11,12* (or *12,11*) quadriplex that is heterozygous in one genome and homozygous in the other genome. Hence, on selfing for even a single generation, either the *11,12* or the *12,11* quadriplexes will produce progeny, one-fourth of which are expected to be heterotic true-breeding *11,22* quadriplexes. Consequently, it is much more likely that novel favorable mutants will survive and become incorporated into tetraploid *Ab* than into *Ah*. Moreover, with the formation of *11,22* quadriplexes in the tetraploid, the stage would be set for development of still additional allelic diversity. Subsequent mutations of allele *1* to allele *3* in frequent, or soon-to-be frequent, *11,22* quadriplexes might have led to quadriplex *22,33* of *Pgd2*. This quadriplex, although less successful than the original *11,11* quadriplex or the highly successful first-step derivative *11,22* quadriplex, clearly found several sites in which it was able to survive (Table 10-1). Several other patterns of quadriplex formation are illustrated in Table 10-1. Three other loci of Table 10-1 (*Mdh3*, *Acp1*, *Acp2*, and usually *Pgi1*) apparently followed a pattern similar to that of *Pdg2*; each formed a single highly successful heteroalleleic quadriplex (in many cases from two alleles not present in the diploid), as well as one or more less successful heteroallelic or homoallelic quadriplexes. Locus *Mdhl* was unique among the 14 loci discussed here; it formed a single heteroallelic quadriplex (*11,22*) that is now ubiquitous and monomorphic worldwide. The intralocus interaction (analogous to true overdominance) between the pairs of homozygous alleles (*11,22*) of this locus clearly led to a level of adaptedness that has withstood all challenges from the many novel mutant alleles that must have appeared over hundreds, or even thousands, of years in the many different environments in which *Ab* has become established worldwide. Another pattern of quadriplex formation (the pattern traditionally postulated for quadriplex formation) is that of the 4 loci (*Est1*, *Lap1*, *Prx1*, and sometimes *Pgi1*) that are highly polymorphic in the diploid. Each of these loci formed several homoallelic and several heteroallelic quadriplexes in *Ab*. Still another pattern of quadriplex formation is that of loci *Pgm1*, *Got2*, *Mdh2* and *Got1*. All of these loci are monomorphic (or very nearly so) in Spain (and worldwide) for a single allele in the diploid as well as for the same allele in the tetraploid. Evidently, these monomorphic alleles encode for some important enzymatic function such that they confer adaptedness in a homoallelic state superior to that conferred by any other alleles or combination of alleles, whether homoallelic or heteroallelic, that have arisen in either *Ah* or *Ab*. Thus, polyploidy provides an opportunity for utilization of intra-allelic interactions among alleles of the same locus, an opportunity that is not available to diploids. Overall, the net effect of polyploidization, followed by diploidization, was a great increase in allelic diversity (38 to 58 alleles, or 37% increase) that occurred in the tetraploid in Spain.

Note also that for most loci the increased allelic diversity provided not only opportunities for favorable intralocus interactions analogous to true overdominance, but even more opportunities for exploitation of favorable epistatic interactions among alleles of different loci (pseudo-overdominance). As an example, consider a diploid plant genotypically *11* for some locus (say *Lap1*) and genotypically *55* for another locus (say *Est1*). Natural heterozygotes are so rare (~1%) in *Ah* that worthwhile exploitation of possible overdominant interactions in 15 duplex heterozygotes seems unlikely. Furthermore, only a single epistatic (pseudo-overdominant) interaction ($11 \times 55$) is possible in *Ah*. However, in a tetraploid population that includes individuals genotypically *22,33* for *Lap1* and *55,77* for *Est1*, 6 interactions are possible, 2 intralocus interactions (analogous to true overdominance) $22 \times 33$ for Lap1 and $55 \times 77$ for Est1, as well as 4 pairwise interlocus epistatic interactions, $22 \times 55$, $22 \times 77$, $33 \times 55$, and $33 \times 77$. It seems significant that all of these combinations are either fixed or very nearly fixed in all 7 populations of the cold Central Northern Plateau of Spain (Figure 10-1), which suggests that all 6 of these particular interactions contribute to adaptedness in the particular habitat of that region. Very large numbers of other nonrandom associations of specific duplexes and quadriplexes within broad geographical regions, and within specific sites within regions, have been observed in Spain and elsewhere. This indicates that some duplexes and quadriplexes confer superior adaptedness over broad regions, whereas some confer superior adaptedness only within local sites within regions.

Very large numbers of different 2-locus, 3-locus, up to 14-locus, epistatic interactions are possible in the diploid and vastly larger numbers of epistatic interactions are possible in the tetraploid. Observed patterns of intralocus and interlocus interactions, as expected, turned out to be much more complex in the tetraploid than in the diploid. Consequently, we first turn to the less-complicated 14-locus genotypes observed in the diploid to obtain some sense of the 2-locus and higher-order epistatic interactions that have been successful in Spain. In total, 107 different 14-locus genotypes were observed in the 10 Spanish populations of diploid *Ah*. The main features of both within-population and among-population differentiation can be deduced from Table 10-2, which gives the frequencies of the single most frequent 14-locus genotype and the total numbers of 14-locus genotypes at each site. First, note that the most frequent 14-locus genotype, presumably the single best-adapted 14-locus genotype at any site, differed from that present in each of the other sites. Only one 14-locus genotype was present in sites 1 and 2; the 14-locus genotype at these two sites was identical at 9 loci but differed at 5 other loci (*Mdh3, Pgi1, Acp1, Prx1, Lap1*). It is, of course, possible that differences in adaptedness between these two sites were due entirely to the main effects of these 5 loci; however, the very large numbers of different patterns in which the alleles of the 14-locus genotypes occur in the diploid suggest that many 2-locus, 3-locus, and still more numerous higher-order interactions possibly also played a role in adaptedness. When discrete log-linear analyses of the data, based on theory developed by Fienberg (1980), were calculated, they revealed that many favorable higher-order multilocus associations of alleles were present in statistically highly significant excess, whereas large numbers of other high-order multilocus associations (presumably unfavorable associations) were either present in statistically significant deficiency or entirely absent. Within-population-diversity in *Ah* varied from none in site 1 and site 2 (only one 14-locus genotype present in each site), to small in site 3 (only two 14-locus genotypes present), to substantial in sites 4 and 5 (8 different 14-locus genotypes present), to high for the 5 remaining sites (13 to 36 different 14-locus genotypes present). Population sizes were relatively large at each site, so that, overall, the effects of genetic drift may have been smaller than the effects of selection. Furthermore, frequent seed exchanges, owing to many different causes, including spread by domestic grazing animals, take place among sites. Such exchanges are expected to dampen the extent of genetic differentiation among sites; hence, it seems likely that the genetic differences that were evident among all sites were not due to genetic drift, but that they resulted primarily from the sorting-out by selection of the single best-adapted 14-locus genotype in the 2 monomorphic sites and the particular mixes of 14-locus genotypes observed in the 8 polymorphic sites of Table 10–2.

Generation-to-generation changes in the frequencies of 14-locus genotypes, especially in response to obvious fluctuations in moisture or temperature stresses, have often been observed in both *Ah* and *Ab*. Year-to-year fluctuations in genetic frequencies are usually small to modest.

Accordingly, the production of novel 14-locus genotypes is almost certainly a slow process in all heavily selfing populations. The reasons were discussed in abstract form in Chapter 5 but are sufficiently important to justify repetition in the light of the data of Tables 10-1, 10-2, and 10-3. The 99% of selfing that occurs in

TABLE 10-2  Most frequent 14-locus genotypes in 10 diploid sites (adapted from Allard et al. 1993)

| | Locus | | | | | | | | | | | | | |
|---|---|---|---|---|---|---|---|---|---|---|---|---|---|---|
| SITE | Pgm1[a] | Got1 | Mdh2 | Mdh3 | Acp2 | Pgd1 | Pgi1 | Acp1 | Prx1 | Lap1 | Est1 | N[b] | fc |
| 1 | 11 | 11 | 11 | 22 | 11 | 11 | 11 | 55 | 55 | 11 | 55 | 1 | 1.00 |
| 2 | 11 | 11 | 11 | 66 | 11 | 11 | 44 | 11 | 22 | 22 | 55 | 1 | 1.00 |
| 3 | 11 | 11 | 11 | 22 | 22 | 11 | 22 | 11 | 11 | 11 | $77^P$ | 2 | 0.93 |
| 4 | 11 | $22^P$ | 11 | 22 | 11 | 11 | $11^P$ | $22^P$ | 22 | $44^P$ | $55^P$ | 8 | 0.71 |
| 5 | 11 | 11 | 11 | 22 | 11 | $11^P$ | $44^P$ | 11 | 22 | $22^P$ | $55^P$ | 8 | 0.51 |
| 6 | 11 | 11 | 11 | 22 | 11 | $55^P$ | 11 | $55^P$ | $44^P$ | $44^P$ | $11^P$ | 13 | 0.45 |
| 7 | 11 | 11 | 11 | 22 | 11 | $22^P$ | $44^P$ | $11^P$ | $22^P$ | $22^P$ | $22^P$ | 20 | 0.26 |
| 8 | 11 | 11 | 11 | $22^P$ | $11^P$ | $11^P$ | $11^P$ | $22^P$ | $11^P$ | $77^P$ | $77^P$ | 17 | 0.21 |
| 9 | 11 | 11 | 11 | 22 | $55^P$ | $11^P$ | $22^P$ | $22^P$ | $22^P$ | 34 | $22^P$ | 34 | 0.17 |
| 10 | 11 | 11 | $11^P$ | 22 | $22^P$ | $22^P$ | $22^P$ | $11^P$ | $11^P$ | $11^P$ | $55^P$ | 36 | 0.12 |

[a]Also *Got2, Mdh1, Pgd2*.
[b]Number of 14-locus genotypes observed at each site.
[c]Frequency within site of most frequent 14-locus genotypes.
[P]Site polymorphic for 14-locus genotypes owing to polymorphism at locus indicated.

TABLE 10-3 Single most-frequent 14-locus genotypes of *Ab* in the Central Northern Plateau (CNP) and the Peripheral Northern Plateau (PNP) of Spain and in four habitats in California (adapted from Allard et al. 1993)

| Region and Site | Pgm1[a] | Mdh1 | Pgd2 | Mdh3 | Acp2 | Pgd1 | Acp1 | Prx1 | Lap1 | Est1 | f[b] |
|---|---|---|---|---|---|---|---|---|---|---|---|
| CNP-43 | 11,11 | 11,22 | 11,22 | 11,22 | 11,44 | 11,33 | 11,22 | 11,11 | 22,33 | 55,77 | 1.0 |
| CNP-45 | 11,11 | 11,22 | 11,22 | 11,22 | 11,44 | 11,33[P] | 11,22 | 11,11[P] | 22,33 | 55,77 | 0.99 |
| CNP-46 | 11,11 | 11,22 | 11,22 | 11,22 | 11,44[P] | 11,33[P] | 11,22 | 11,11 | 22,33 | 55,77[P] | 0.68 |
| CNP-33 | 11,11 | 11,22 | 11,22[P] | 11,22[P] | 11,44[P] | 11,33[P] | 11,22 | 11,11[P] | 22,33[P] | 55,77[P] | 0.91 |
| PNP-1 | 11,11 | 11,22 | 11,22[P] | 11,22[P] | 33,44[P] | 11,33[P] | 11,22[P] | 11,11[P] | 22,33[P] | 55,77[P] | 0.48 |
| PNP-3 | 11,11 | 11,22 | 11,22[P] | 11,22[P] | 11,44[P] | 11,33[P] | 11,11[P] | 11,22[P] | 22,33 | 55,77[P] | 0.33 |
|  |  |  |  |  |  |  |  |  |  |  | rDNA |
| CA-M[c] | 11,11 | 11,22 | 11,22 | 11,22 | 33,44 | 22,22 | 11,33 | 11,22 | 11,33 | 55,55 | 15,9,8,7 |
| CA-X[c] | 11,11 | 11,22 | 11,22 | 11,22 | 33,44 | 11,11 | 11,22 | 11,33 | 11,22 | 33,55 | 13,10,8,7 |
| CA-H[c] | 11,11 | 11,22 | 11,22 | 11,22 | 33,44 | 11,22 | 11,22 | 11,33 | 11,22 | 33,35 | 10,8,7 |
| CA-JR[c] | 11,11 | 11,22 | 11,22 | 11,22 | 33,44 | 11,22 | 11,22 | 11,22 | 11,22 | 33,55 | 12,8,7 |

[a] Also *Got2*, *Got1*, *Pgi1*.
[b] Frequency within site of single most frequent 14-locus genotype.
P Locus polymorphic within site.
[c] Polymorphic in areas where M, X, H, and/or JR habitats form interfaces with each other.

*Ah* (and in *Ab*) quickly forces all loci to complete or to nearly complete homozygosity, with the result that almost no effective recombination takes place after the third generation of selfing, even for loci located on different chromosomes. At the same time the ~1% of outcrossing that occurs leads to short bursts of segregation and recombination in the first three generations after an outcross. Segregation and recombination are expected to produce many novel multilocus genotypes, some of which may be superior to existing genotypes. If an outcross $F_1$ plant is heterozygous for *N* loci, only half of its selfed progeny are expected to be heterozygous in the next generation, and thereafter selfing will reduce heterozygosity by 1/2 in each succeeding generation. Thus, the series for reduction of heterozygosity within each of the lineages descended from any single outcross individual will be 1/2, 1/4, 1/8, 1/16, so that >90% of the total recombination will have occurred in the first four generations following an outcross. Each lineage would soon be expected to approach an equilibrium featuring 2*N* (possibly many more) equally frequent homozygous lines. However, selection is expected to favor the more fit and discriminate against the less fit multilocus genotypes, and, under the highly disciplined "line breeding" that occurs with selfing, favorable alleles and favorable combinations of alleles may more often be expected to be preserved than the original genotypes in the population.

At the same time, populations of *Ah* and *Ab* are often dense and plant-to-plant competition is high, so that each plant produces few seeds. Hence, the number of plants within each lineage is likely to be small, with the result that many fit genotypes would almost certainly drift out of each lineage, whereas some of the less-fit genotypes might survive by accident. Thus, the rate of genetic improvement per generation is unlikely to be remarkably large. Regardless, the survivors will quickly be driven to homozygosity by selfing, and all loci within each lineage, whether located on the same or different chromosomes, will behave thereafter as if they are tightly linked with recombination value ~0.01 (see Chapter 5). Hence, any two loci (say *a* and *b*), no matter where they are located within the genome, will behave as a single functional locus, *c*, and they will manifest some degree of pseudo-dominance up to a level of pseudo-overdominance whenever $c_1 = a_1b_2, c_2 = a_2b_1$ and the survival value of $C_1C_2$ is < $C_1C_1\ C_2C_2$. Selfing is thus a simple, straightforward, and highly effective way of concentrating favorable alleles within inbred lines, as well as for exploiting many favorable epistatic interactions among alleles of different loci, whether or not the loci are located on the same or different chromosomes. Thus, on one hand, the ~1% of outcrossing that occurs in most generations provides a small but steady supply of often multiply heterozygous natural outcrosses that segregate to produce numerous novel multilocus genotypes, some of which may be superior. On the other hand, however, the 99% of selfing restricts recombination sufficiently to protect virtually all previously existing, as well as virtually all newly arisen, favorable epistatic combinations of alleles from destruction by segregation. This raises an important question: At what level of outcrossing is the production of favorable new epistatic combinations of alleles counterbalanced by the destructive effects of segregation on favorable epistatic combinations of alleles? Both *Ah* and *Ab* are very difficult to hybridize artificially and, hence, neither is favorable material for addressing this question experimentally. As will be discussed in Chapter 11, experiments with barley, a heavily selfing cultivated species in which hybrids can be made relatively easily, show that increasing outcrossing rates artificially from the normal 1% to as little as 5% immediately leads to dramatic alterations in population structure, accompanied by large decreases in seed yields that persist for many generations after selfing is allowed to return to the normal levels of ~1%. Thus, the effects of even a single generation of increase in the rate of outcrossing is likely to be devastating in inbreeding populations made up of many different genotypes. However, if such populations are allowed to resume their normal level of inbreeding, their performance often continues to improve for several generations, presumably because of development of new favorable multiallelic combinations through the line breeding that occurs automatically as a result of the return to ~99% of selfing.

Returning to *Ab*, the number of 14-locus genotypes observed in the 50 Spanish populations of Figure 10-1 exceeded 440, about four times more than were observed in the 10 populations of *Ah*. Fourteen-locus genetic structure featuring interlocus epistasis was clearly much more complex in *Ab* than in *Ah*, and it was also much more closely attuned to specific environments. This can be illustrated by description of the very close associations that developed between certain specific 14-locus genotypes of *Ab* and particular habitats, as well as between various distinctive habitats and particular polymorphic mixtures of 14-locus genotypes. Among many such associations, those between the genotypes and the cold, high-elevation habitats of the Central Northern Plateau (C.N.P., Figure 10-1) were especially distinctive. A specific 14-locus genotype was the most frequent genotype in all 7 sites of that region (sites 43–48, 33); this particular genotype was monomorphic ($F = 1.00$) in site 43 and nearly monomorphic ($F = 0.99$) in site 45, the two highest and coldest sites in which *Ab* is able to survive in the Northern Plateau Region. Two changes took place in the 5 lower elevation sites of the C.N.P.: the frequency of the "cold-tolerant" 14-locus genotype fell off sharply (as in site 46); and/or several additional loci became polymorphic (as in site 33) in these 5 lower sites. The cold-tolerant genotype was entirely absent in all sites located in still lower and less cold sites peripheral to the C.N.P. For example, in site 1 of the Peripheral Northern Plateau (P.N.P.-1, Figure 10-1), the cold-tolerant genotype was completely supplanted by a 14-locus genotype that differed at only 1 locus (i.e., quadriplex *11,44* of *Acp2* was replaced by quadriplex *33,44*), whereas in site P.N.P.-3 the cold-tolerant genotype was supplanted by a 14-locus genotype that also differed at only 1 locus (i.e., quadriplex *11,11* of *Prx1* was replaced by quadriplex 11,22). At the same time, these two sites (sites P.N.P.-1 and P.N.P.-3) also become polymorphic for several other loci, indicating that substantial changes had occurred in the patterns of multilocus epistatic interactions. Overall, progressively larger environmental changes were accompanied consistently by progressively greater restructuring of the 14-locus genotypic makeup of the populations. Evidently, natural selection not only sorted out the single 14-locus genotype that is best adapted in the very coldest areas, but also restructured different combinations of 14-locus genotypes into complex mixtures of genotypes that apparently provide superior adaptedness in the several less cold environments of the lower elevation sites. It is noteworthy, however, that among the 50 Spanish populations of *Ab* of this study, only 4 (8%) were monomorphic for a single 14-locus genotype, whereas more than 20 different 14-locus genotypes were present in the great majority (44/48 = 92%)

of the populations. Results such as these indicate that superior adaptedness in *Ah* and *Ab*, as in all other heavily selfing populations that have been studied in detail, has nearly always been associated with extensive genetic diversity, featuring complex mixtures of highly fit nonsegregating multilocus genotypes. Clearly, epistatic interallelic interactions are common not only at the 2-locus level but at all levels up to and including the 14-locus level and, presumably, still higher levels. Heavily selfing populations are rarely monogenotypic. They are clearly not "stuck" in or limited to evolutionary blind alleys of rigid monogenotypic homozygosity, as has been proposed in most treatises on evolution. To the contrary, their genetic systems appear to be exceptionally well equipped to generate and amalgamate complex blends of stable and superior homozygous genotypes into functionally superior mixes that provide close adaptedness in each of the environmental patches they occupy.

## Colonial Populations of Ab

The 14-locus allozyme genotypes and the 2-locus rDNA genotypes in the four major habitats in which genotype-habitat associations have been established to date for colonial populations of *Ab* in California (M = Mesic, X = Xeric, H = Hopland, JR = Jasper Ridge). Present day ancestral Spanish and colonial Californian populations are very closely similar in allelic composition when compared on a locus-by-locus basis: the predominant alleles of Spain are also the predominant alleles in the colonial populations. However, the ancestral and colonial populations differ in genotypic configurations in two main ways. First, the alleles that are predominant in Spain often occur in very different combinations within quadriplexes in California. Second, and more striking, the 14-locus genotypes of California are entirely different from those of Spain. Another difference is that few of the rare alleles of Spain occur in California. The rare alleles in the Californian populations are almost entirely different from those of Spain; thus, the rare alleles present in California are, apparently, nearly always descended from novel mutations that must have occurred in the 100 to 150 generations or so in which *Ab* has been widespread in California. Adaptedness in the colonial populations, as well as in the ancestral Spanish populations, clearly depends on the same set of frequent alleles and rarely, if ever, on rare alleles. Consequently, it appears that changes in allelic frequencies, which are often considered to constitute the elementary process of evolution, played, at most, a minor role in the adaptive changes that occurred in the colonial Californian populations. Instead, a different process—namely, reorganization of the predominant alleles of the ancestral Spanish populations into novel, stable, homozygous, and complexly integrated combinations adapted to specific habitats in California— played the major role in the evolution of the colonial populations. This dramatic reorganization, especially apparent with allozyme loci *Acp2*, *Pgd1*, *Acp1*, *Prx1*, *Lap1*, and *Est1* obviously occurred quite quickly, because *Ab* has been widespread in California for no more than 100 to 150 generations, a very short time in an evolutionary context. The process of adaptation to Californian environments clearly continues to the present. Many Californian populations remain highly polymorphic, especially in areas in which different habitats adjoin one another and frequent hybridizations occur among genetically different incipient ecotypes. Only interactions among nuclear loci have been discussed in this chapter. However, intricate interactions between and among nuclear loci and chloroplast DNA geno-types have also been documented in *Avena* (*Ah, Ab*), as well as in barley and other crop species. The functional basis at the physiological level of these very complex interaction systems is obscure (each marker locus probably identifies a short chromosome segment carrying many genes) and, hence, is likely to remain obscure. However, the genetic evidence leaves little doubt concerning the reality or the importance of such interaction systems in the evolution of adaptedness in plants.

Only allozyme and rDNA variants governed by loci of the nuclear genome have been discussed here. However, it is important to note that interactions have also been reported in various wild species among loci of the nuclear and the chloroplast genomes (e.g., Saghai Maroof et al. 1984). This suggests that plant breeders ulti-mately may also have to take intergenomic interactions into account if breeding operations are to be conducted on an entirely rational basis.

# ELEVEN

# *Marker-Assisted Dissection of Adaptedness in Cultivation*

Modern cultivars are better adapted and much more productive in agricultural environments than their wild ancestors and early domesticates. Although adaptations to the great variety of habitats that exist in nature are complexly inherited traits (Chapter 10), the inheritance of adaptedness is almost certainly no less complex in cultivation. In addition to the ability to carry out essential biological processes efficiently, successful cultivars must also satisfy numerous production requirements of farmers and must meet various quality preferences of the market-place. Landraces of major crops perform agricultural tasks better than their wild ancestors, early domesticates and modern cultivars are often conspicuously more successful, at least in high-input agricultures, than the landraces they replaced. Figure 5-1 corroborates the latter assertion: The average yield of corn in the United States was ~25 bushels per acre from the 1860s through the 1930s, but by the 1990s yields of corn were more than five times greater. During the latest 70-year period of mass-selected open-pollinated varieties of corn (1860–1930) the rate of gain in yield was about 1/50 bushel/acre/year. During the ~30-year period of double-cross hybrids the rate of gain increased to slightly more than one bushel/acre/year (50-fold higher), and in the 25-year period of single-cross hybrids the annual rate of gain rose to nearly 1.76 bushels/acre/year. The increases in rate of gain in yield were due partly to the superiority of hybrid varieties and partly to better management practices (e.g., more fertilizer, better cultural practices). A number of studies have shown that somewhat more than half of the increase in corn yield since 1930 was attributable to genetic improvement. Some now hold the opinion that gains from superior management practices with corn may be fast approaching a cost-gain barrier and that further gains are less likely to be due to superior management practices than to further genetic improvements.

Plant-breeding progress, regardless of crop, depends on choosing appropriate germplasm and on applying efficient breeding methods to the germplasm selected. There can be no doubt that significant progress has been made in improving the rate of gain in the yielding ability of many crops. Yet, the small rate of gain apparent in Figure 5-1 makes clear that the mass selection breeding procedures in corn that prevailed from about 1860 to 1930 were much less effective than the double-cross hybridization methods of the period from about 1930 to about 1970. Double-cross methods were also not fully efficient; for instance, a distressingly small percentage of the thousands or even millions of inbred lines that were developed in the period 1930–1970 did not advance even to test crossing and far fewer, perhaps fewer than 1/10,000, were used to any significant extent in actual commercial agricultural production. Furthermore, as shown by Figure 5-1, single-cross-breeding method-ology led to rates of gain nearly twice those of double-cross methodology. It therefore seems appropriate to ask whether methods of choosing germplasm, the breeding approaches applied, or both were faulty prior to 1930–1970. Although the methods of choosing germplasm and the breeding methods that were applied to the selected germplasm were obviously both successful to an extent, more rapid recent advances suggest that opportunities still exist to improve methods of both choosing the germplasm and developing breeding and selection procedures that more efficiently identify and combine favorable alleles into superior genotypes, not only in corn, but also in many other crops.

In recent years it has been found that examination of the changes that occurred over generations in the frequencies of discretely distinguishable alleles, and particu-larly in the frequencies of discretely distinguishable multilocus genotypes, allow insightful deductions concerning the genetic mechanisms that have led to improved adaptedness and productivity in cultivated plants. In a number of cases that have been examined in detail it has been found that observed evolutionary changes are due to selection acting on several to many short chromosomal segments in which individually distinguishable marker loci reside. It is immaterial to breeders and farmers whether the gain is due to the marker loci themselves or to other loci, including loci that affect quantitative characters, that occupy the same short chro-mosome segment as the marker locus. The outcome of such selection has the advantage that it can be characterized in terms of distinct genetic entities. Marker-assisted analyses of complexly inherited traits have become feasible, and such analyses have increasingly supplemented procedures such as taking measurements of yielding ability under a range of environmental conditions.

We now turn to the main purpose of this chapter, to examine the genetic changes that have occurred in various crop species in response to the very large numbers of generations of selection that have been practiced for high and steady performance in the successions of agricultural environments in which agricultural plants have been grown both during domestication and cultivation in various places in the world. Genetic change will be examined at two levels: first, at the level of frequency changes that occurred in single discretely identifiable marker alleles, and then in terms of complexes of marker alleles of two or more different loci that have often led to high adaptedness in specific environments. In most of the cases to be considered the domesticators and farmers who were involved could not have been aware of the existence of marker alleles themselves, because the technology required to identify such alleles has been developed in only the last two decades or so. Hence, artificial selection could hardly have been directed at either the individual marker alleles themselves or at multilocus complexes of markers. It is likely, however, that any adaptively favorable alleles of other loci located within the same short chromo-some segment occupied by a marker locus would drag some favorable markers along, thus enhancing the chances of survival of the markers. The importance of marker loci is hence solely that they identify short chromosome segments that carry a preponderance of favorable alleles. Moreover, such chromosome segments must be sufficiently small to remain intact through many cycles of crossing over and segregation. Illustrative genetic changes that have occurred over many generations in populations of inbreeders will be documented with data from barley, oats, and rice, whereas the genetic changes that have occurred over time in outbreeders will be documented with data from corn.

## ALLELIC DIVERSITY IN WILD AND CULTIVATED BARLEYS

Table 11-1 gives the numbers of alleles observed at 20 or so marker loci in a number of present-day wild populations of diploid *Hordeum vulgare*, ssp. *spontaneum* (abbreviated *Hs*) from the Middle East. *Hs* is morphologically closely similar to cultivated *Hordeum vulgare*, and it is widely accepted as the wild progenitor of cultivated barley. The overall genetic structures of all wild barleys are also closely

TABLE 11-1  Numbers of alleles of commonly studied loci encountered in two samples of wild barley (*Hs*) and cultivated barley (*Hv*)

| | NO. OF LOCI | NO. OF ALLELES | MEAN NO. OF ALLELES/LOCUS | RELATIVE NO. OF ALLELES/LOCUS |
|---|---|---|---|---|
| | | Sample A[*] | | |
| *Hs* (Israel) | 19 | 77 | 4.05 | 100 |
| *Hv* (12 Iranian landraces) | 25 | 56 | 2.24 | 55 |
| *Hv* composite cross 21 ($F_{17}$) | 20 | 34 | 1.70 | 42 |
| *Hv* composite cross 34 ($F_4$) | 20 | 36 | 1.80 | 44 |
| | | Sample B[**] | | |
| *Hs* (Middle East)[†] | 20 | 103 | 5.15 | 100 |
| *Hv* (Middle East)[‡] | 20 | 55 | 2.75 | 53 |
| *Hv* (CCII, $F_7$–$F_9$) | 26 | 41 | 1.58 | 31 |
| *Hv* (CCII, $F_{53}$) | 26 | 37 | 1.42 | 28 |
| *Hv* (CCV, $F_6$) | 25 | 39 | 1.56 | 31 |
| *Hv* (CCV, $F_{35}$) | 25 | 36 | 1.44 | 28 |
| *Hv* (CCXXI, $F_4$) | 25 | 42 | 1.68 | 33 |
| *Hv* (CCXXI, $F_{22}$) | 25 | 38 | 1.52 | 30 |
| *Hv* (9 California cultivars)[§] | 25 | 36 | 1.44 | 28 |

[*] Sample A: Brown and Munday 1982. [**] Sample B: Kahler and Allard 1981; Kahler et al. 1981; Saghai Maroof et al. 1984, 1990; Allard (unpublished data). Sample A included only allozyme loci. Sample B included loci affecting allozyme and restriction fragment variants in *Hs* and allozyme, restriction fragment, and morphological variants in the composite crosses.

[†] Israel, Lebanon, Syria, Jordan, Turkey, Iran, Afghanistan.

[‡] Landraces from Lebanon, Syria, Jordan, Turkey, and Iran.

[§] Mean number of alleles/cultivar was 27.2 (many cultivars were polymorphic, especially for *Est* loci *1*, *3*, and *4*).

similar to those of diploid *Ah*, described in some detail in Chapter 10, as well as closely similar to those of cultivated barleys and other highly selfing diploid species. Table 11-1 gives the number of alleles found in two samples of wild barley and in several samples of cultivated barley (*H. vulgare, ssp. vulgare*, abbreviated *Hv*) in various stages of domesticity. Almost all loci examined (~99%) in wild barley populations have been found to be homozygous, usually for the same allele in all populations. But different individuals within wild barley populations were often homozygous for two different alleles of the same locus (one allele was usually much more frequent in each population than the other allele), whereas different individuals in other populations were fixed, or nearly fixed, for three different alleles of the same locus (again, one allele was usually much more frequent than either of the other alleles). The average number of alleles per locus was ~4 in each of the wild populations of Table 11-1. The total number of multilocus genotypes per population was, nevertheless, quite large in most wild populations of *Hs* owing to the numerous multilocus combinations of alleles that can be generated by several loci, each with only two alleles, for example, $2^2 = 4$ different 2-locus homozygotes (*AABB, AAbb, aaBB, aabb*) possible with only 2 diallelic loci, $2^3 = 8$ different 3-locus homozygotes with 3 triallelic loci, and $2^4 = 16$ different homozygotes with 4 alleles per locus. However, wild, as well as cultivated, populations of barley examined have usually been dominated by only a very few homozygous multilocus genotypes, these genotypes composed nearly exclusively of the most frequent (predominant) alleles. Table 11-1 also gives the allelic composition of several different populations of cultivated barleys in various stages of domesticity. The main conclusion drawn from studies of such changes is that the numbers of alleles of the 25 or so marker loci monitored per population decreased sharply in the progression from the wild populations to modern landraces; however, the decrease became much slower later in the process as modern cultivars were developed. Adaptedness and productivity in agricultural environments also increased substantially during each of the steps in the progression from wild to modern agricultural conditions. This leads to an important question: Which alleles of wild barley survived, and which alleles were eliminated, during domestication and during many generations in cultivation in many different habitats throughout the world? By far the most common pattern was that only those few alleles that were present in *Hs* in the highest frequencies nearly always survived through hundreds of generations of natural selection and/or human-directed selection in cultivation, whereas the less frequent and, especially, the rare alleles of *Hs* were nearly always eliminated. Concomitantly, the number of homozygous-multilocus genotypes present in modern cultivars also decreased relative to the numbers that had been present in wild ancestral populations. Moreover, the relatively few surviving multilocus genotypes were composed almost entirely of homozygotes made up of alleles that had been frequent in *Hs*. This suggests that selection over the centuries had sorted out many short chromosome segments, not surprisingly those marked by frequent alleles, and that those segments also contributed to superior survival ability and performance under agricultural conditions. The homozygosity of self-pollinating species was apparently important, because homozygosity largely forestalls the breakup by segregation of superior homozygous segments of chromosomes when occasional outcrosses between individuals with different genotypes occur in the cultivated populations. At the same time the occasional outcrosses that occur, when followed by selfing, lead to rigorous "line breeding" that encourages the development and the incorporation of additional favorable epistatic combinations of alleles of different loci into the populations, as well as elimination of alleles and short chromosome segments with poorer combining ability (see the last section of Chapter 5).

## THE POPULATION DYNAMICS OF ALLELIC FREQUENCY CHANGE

A feature that became increasingly apparent in early studies of Mendelian inheritance in barley is that patterns of allelic frequency change in segregating families are often useful in predicting the probable ultimate worth of alleles. This can be illustrated with studies conducted in the 1960s on $F_2$ families in which the predominant allele, allele *112*, of ribosomal DNA locus *Rn1* and any one of the less-frequent alleles of this locus (Table 11-2) were segregating. The expected 1:2:1 ratios in such segregations were consistently distorted; allele *112* was regularly present in great excess in generation $F_2$, whereas all of the rare or infrequent alleles of this locus were consistently in deficiency. Thus, even under the conditions of little or no plant-to-plant competition under which the inheritance studies were conducted, short chromosome segments marked by rare and infrequent alleles were inferior in survival ability as compared with short chromosome segments marked by frequent alleles. Moreover, when populations that had been started by bulking $F_2$ seeds of hybrids between wild barleys and cultivated barleys were advanced into $F_3$, $F_4$, and later generations that were grown in cultivation under conditions that led to moderate plant-to-plant competition, the advantage of chromosome segments marked by frequent alleles over segments marked by less-frequent alleles became even more striking. The frequent wild-type alleles rapidly increased in frequency, and the less-frequent wild-type alleles either rapidly decreased in frequency or were quickly eliminated in later generations.

Only 2 among the 12 alleles of locus *Rrn1* found in wild barleys were found in a sample of 1,032 cultivated barleys that had been chosen by random methods from the worldwide barley collection of the U.S. Department of Agriculture (USDA) (Table 11-2). These were allele *112* (F = 0.98), the most frequent allele in wild populations, and allele *111* (f = 0.02), the second most frequent allele in wild populations. Only 3 alleles of locus *Rrn1* (Table 11-3) were found in another sample of cultivated barleys, the 28 barley varieties chosen by Harlan and Martini (1929) to represent the best-available cultivated germplasm of the major barley-growing areas of the world in Composite Cross II (CCII). These were alleles *112* (F = 0.86), *111* (F = 0.07) and *110* (F = 0.07), which were, respectively, the three most frequent alleles of locus *Rrn1* in wild barleys. However, by generation $F_8$ of CCII, the short chromosome segment marked by allele *112* had become virtually monomorphic in cultivated barleys, whereas the short chromosome segments marked by alleles *110* and *111* had ben reduced to low frequency, only to disappear completely by generation $F_{23}$ (Table 11-3). This indicates that the chromosome segments marked by allele *112* made more useful contributions to survival, and presumably also to adaptedness, in cultivation than any among the other 11 chromosome segments

TABLE 11-2   Alleles of rDNA loci *Rrn1* and *Rrn2* in wild barley (*Hs*) from Israel and Iran and from a worldwide sample of cultivated barley (*Hv*) (from Saghai Maroof et al. 1990)

| LOCUS AND ALLELE | WILD BARLEY | | CULTIVATED BARLEY | |
|---|---|---|---|---|
| | No. | Freq. | No. | Freq. |
| *Rrn1* | | | | |
| *108a* | 29 | 0.06 | 0 | 0.00 |
| *105* | 100 | 0.19 | 0 | 0.00 |
| *109* | 68 | 0.13 | 0 | 0.00 |
| *110* | 10 | 0.02 | 3 | 0.02 |
| *111* | 10 | 0.02 | 3 | 0.02 |
| *112* | 273 | 0.53 | 187 | 0.98 |
| *113* | 4 | < 0.01 | 0 | 0.00 |
| *114* | 3 | < 0.01 | 0 | 0.00 |
| *115* | 0 | 0.00 | 0 | 0.00 |
| *116* | 2 | < 0.01 | 0 | 0.00 |
| *117* | 1 | < 0.01 | 0 | 0.00 |
| *118* | 0 | 0.00 | 0 | 0.00 |
| Total | 518 | 1.00 | 190 | 1.00 |
| *Rrn2* | | | | |
| *100* | 6 | 0.01 | 2 | 0.01 |
| *101* | 0 | 0.00 | 3 | 0.02 |
| *102* | 4 | < 0.01 | 2 | 0.01 |
| *103* | 4 | < 0.01 | 0 | 0.00 |
| *104* | 13 | 0.02 | 116 | 0.65 |
| *105* | 25 | 0.05 | 0 | 0.00 |
| *106* | 69 | 0.13 | 2 | 0.01 |
| *107* | 427 | 0.78 | 53 | 0.30 |
| Total | 548 | 1.00 | 178 | 1.00 |

marked by alleles of *Rrn1*. It is tempting to regard allele *112* itself as superior to the 11 other alleles of locus *Rrn1*, but, as we will see in the next section of this chapter, it turned out to be difficult to resolve this issue experimentally.

Another unusually polymorphic ribosomal DNA locus, *Rrn2*, illustrates a very different and quite unusual pattern of allelic frequency change in barley (Table 11-3). Although allele *107* of this locus was by far the most frequent (F = 0.78) among 8 alleles of *Rrn2* in wild barley (*Hs*), allele *107* was much less frequent (F = 0.26) in the cultivated parents of CCII than allele *104*. Nevertheless, allele *107* thereafter steadily increased under California conditions, whereas allele *104* steadily decreased in frequency until these two alleles appeared to approach equilibrium values of 0.70 and 0.30, respectively, after 53 generations. Despite substantial fluctuations in frequency, allelic-frequency changes ultimately identified alleles *107*

TABLE 11-3 Frequencies of alleles of rDNA loci *Rrn1* and *Rrn2* in wild barley (*Hs*) and in the 28 parents and various generations of Composite Cross II (CCII). Alleles present in frequency < 0.01 in CCII are not listed (from Saghai Maroof et al. 1984, 1990).

| | | | | GENERATION OF CCII | | | |
| ALLELE | WILD BARLEY | PARENTS | $F_8$ | $F_{13}$ | $F_{23}$ | $F_{45}$ | $F_{53}$ |
|---|---|---|---|---|---|---|---|
| | | | | *Rrn1* | | | |
| 110 | 0.05 | 0.07 | | 0.01 | | | |
| 111 | 0.02 | 0.07 | 0.02 | | | | |
| 112 | 0.53 | 0.86 | 0.98 | 0.99 | 1.00 | 1.00 | 1.00 |
| | | | | *Rrn2* | | | |
| 101 | 0.00 | 0.02 | | | | | |
| 102 | < 0.01 | 0.02 | | | | | |
| 104 | 0.02 | 0.66 | 0.49 | 0.48 | 0.41 | 0.46 | 0.30 |
| 106 | 0.13 | 0.04 | 0.01 | | 0.02 | | |
| 107 | 0.78 | 0.26 | 0.50 | 0.52 | 0.57 | 0.54 | 0.70 |

and *104* as the 2 most useful among the 8 alleles of *Rrn2*. About two-thirds of modern elite Californian varieties are homozygous for allele *107* and about one-third are homozygous for allele *104*; this suggests that the short chromosome segment marked by allele *107* has been favored slightly in cultivation in California relative to the chromosome segment marked by allele *104*, and that these two alleles are greatly favored relative to all other alleles of this locus.

The brittle (shattering) versus nonbrittle (tough) rachis traits discussed in Chapter 3 provide an example of control by alleles that are monomorphic in the wild but that are not present in cultivation. The closely linked dominant allelic pair *Btl-Bt2*, fixed in *Hs*, together produce the brittle rachis trait that confers high adaptedness as a seed-dispersal mechanism in wild barley, whereas the recessive pair *btl-bt2* presumably confer high adaptedness in cultivation as a harvest aid. Selection apparently reversed direction when barley was domesticated (the archaeological record makes it clear that barley became nonshattering, presumably as a harvest aid, very shortly after the species was taken into domestication).

The alleles of barley, and of other self-pollinated species that have been studied in detail, appear to fit into three somewhat overlapping classes: (1) alleles that are predominant in a high proportion of cultivated varieties and breeding stocks representing a broad range of environments; (2) alleles that are present in intermediate-to-high frequencies in the varieties and breeding stocks of some, but not all, ecogeographical regions; and (3) alleles that are absent or infrequent in the varieties and breeding stocks of nearly all ecogeographical regions. The alleles of Category 1 appear to make significant contributions to wide adaptedness and high productivity in nearly all genetic backgrounds and in nearly all environments. The population biology of alleles of Category 2 suggests that their contributions to adaptedness and performance often differ widely in different ecogeographical regions and in different genetic backgrounds, as well as from year to year. Data from both wild and cultivated populations indicate that alleles of this category often ultimately enter into favorable interlocus epistatic interactions in diploids, and perhaps even more often enter into favorable intralocus interactions in polyploids. The frequently localized distributions of such alleles suggests that they should be given some priority in the sampling of germplasm; it is possible that useful alleles of this sort are still to be identified. However, the population biology of alleles of Category 3 suggests that they are nearly always defective in some way, or ways, in nature as well as in cultivation. Thus, infrequent alleles usually appear to have little of value to offer in crop improvement programs.

In some instances changes in allelic frequencies over generations have not been monotonic in barley, but have often shifted in amount of change and/or in the direction of change from year to year. A number of analyses of Californian barley populations suggest that such short-term shifts in frequency are often associated with short-term shifts in environment, usually with variations in rainfall. Such results led to the conclusion that changes in allelic frequencies often provide quick and inexpensive, but useful, measures of survival values and adaptive properties of the individual alleles themselves and/or the short chromosome segments they mark. Improved efficiency in testing for fitness to specific features of the environment has been achieved frequently by identifying pairs of conveniently located test areas that differ sharply in relevant environmental characteristics (e.g., usually ample moisture vs. moisture stress, fertile vs. infertile vs. soil). Growing the same populations in such contrasting environments for only year or two has often identified the environment or environments in which particular alleles prospered versus those in which they did not prosper.

## PLEIOTROPY VERSUS LINKAGE

The closeness of the associations between many specific alleles and adaptedness suggested that the marker alleles themselves might possibly play significant roles in adaptive processes. A program designed to test this hypothesis was consequently initiated with a number of easily classified characters of the spike of barley (e.g., two-row vs. six-row, rough vs. smooth awn, black vs. light lemma color), each a characteristic that has been long regarded as controlled by a single major allele of a single locus. The effects of such distinguishable alleles on various quantitative traits were determined by developing numerous pairs of highly isogenic lines, either by repeated backcrosses to one or the other or both parents or, more often, by recombinant-inbred line procedures. Comparisons between alternative homozygotes extracted after many generations (usually 15 to 20 generations of enforced heterozygosity for each marker locus) revealed that each such allele monitored had precisely the same effect on quantitative characters following multiple opportunities for recombination. It was concluded that if other loci affecting quantitative characters were located in the same chromosome segment as the marker loci, the marked segment of chromosome in nearly all cases must have been very short, so short that little or no crossing over had occurred in any instance within the marked segment.

## ANALYSES OF MULTILOCUS COMBINATIONS

Although precise locus-by-locus analyses of the consequences of combining plural discrete genetic loci into multiplex rearrangements pose many practical problems, the main features of interactions involving only a few loci segregating simultaneously can sometimes be resolved by simple plots of frequency changes of the multiple gametic types that normally occur over a number of generations. Figure 11-1 gives the changes that occurred in 4-locus gametic frequencies of 4 *Esterase* loci (*Est1, Est2, Est3, Est4*) in barley CCII that had been grown at Davis, California, for 53 generations and at Bozeman, Montana, from generation $F_{26}$ to $F_{53}$. *Esterase* loci *1, 2,* and *3* are very tightly linked on chromosome 3 of barley (the crossover value between loci *1* and *2* is 0.0023, that between loci *2* and *3* is 0.0042, and that between loci *1* and *3* is 0.0059), whereas locus *Est4* is located on chromosome *7*. Consequently, owing to such very close linkages, the alleles of loci *1, 2,* and *3* must nearly always be transmitted as an intact group of three alleles, whereas the alleles of locus *Est4* are expected to segregate independently of those of loci *1, 2,* and *3*. Although 13 different gametic types were found among the 16 most likely gametic types in the early generations ($F_2$ to $F_6$) of CCII, only two gametic types, *1221* and *2112*, increased significantly in frequency at Davis over generations. Thereafter, at Davis, gametic type *2112* declined in frequency from generations $F_7$ to $F_{20}$ (a series of very dry years), but it increased steadily in frequency thereafter (generally years of ample rainfall) to F = 0.95 by generation $F_{53}$. In contrast, gametic type *1221*, after an early increase in frequency during drought years declined steadily in frequency during the years of ample rainfall, to F = 0.02 by generation $F_{53}$. Contrariwise, either gametic type *2112* or, perhaps, alleles of other undetected but different loci located within the two small chromosome blocks that are marked by *Esterase* loci *1, 2, 3,* vs. *Est4* are clearly favored in wet years, whereas gametic type *2112* is at a disadvantage in dry years.

When the Davis population was in generation $F_{26}$, a large sample of seeds was sent to Bozeman, Montana, where E. S. Hockett grew the population, without conscious selection, for 19 generations ($F_{27}$ to $F_{46}$). In the continental climate of Montana gametic type *2112* (the undoubted winner in California in wet years) declined steadily in frequency (Figure 11-1), whereas gametic type *1221* (the undoubted winner in California in dry years) did even more poorly; it had nearly disappeared from the Montana population by generation $F_{46}$. In Montana crossover gametic type *2121* increased slowly but steadily in frequency from F = 0.10 to about F = 0.21, whereas crossover gametic types *1212* and *1112* increased slowly in frequency from about generation 25 to generation 33 but rather rapidly from generation 33 to generation 46. Although it was not clear as to which, if any, of these three gametic types might ultimately have become predominant in the North Plains States or the Prairie Provinces of Canada, it seems significant that gametic type *1212* is the single most frequent gametic type observed in the commercial barley varieties of that large and environmentally diverse ecogeographical region. Later, a particularly informative study of the four esterase loci was conducted on a Davis population that was descended from a single $F_1$ hybrid plant obtained by crossing the two least well adapted parents of CCII, one of which carries gametic type *1122* and the other gametic type *2211*; recall that the favored types in California are *1221* and

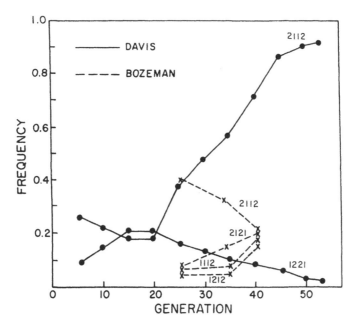

FIGURE 11-1  Multilocus gametic frequencies for *Est1*, *Est2*, *Est3*, and *Est4* in various generations of CCII grown in Davis, California, and Bozeman, Montana.

*2112*. In generation $F_4$ of the *1122 × 2211* cross, rare single-crossover gametic types were detected between the tightly linked loci *1*, *2*, and *3*. By generation $F_9$ double-crossover gametes had appeared, by generation $F_{21}$ gametic type *2112* had become the single most frequent gametic type in the population, and by generation $F_{45}$, gametic type *2112* had reached frequency f = 0.83. Clearly, despite the very close linkages between loci *2* and *3*, segregation and recombination had broken down the two original inferior *1122* and *2211* 4-locus configurations to produce the clearly superior *2112* configuration, which natural selection thereafter rapidly increased to frequency >0.90 by generation $F_{46}$.

Another feature of evolutionary change of apparent significance to plant breeders in CCII and other populations is that the total number of loci involved in favorable interlocus interactions increased over generations. In early generations most of the 20 or so electrophoretically detectable loci monitored were unassociated except for the 4 *Esterase* loci that made up one exceptionally tightly knit group. However, within a few more generations certain among these 20 or so other loci had coalesced into 2, 3, or 4 clusters of 4 or 5 loci each. This suggested that assembly of larger clusters of alleles had increasingly enhanced adaptedness and that, given sufficiently large numbers of marker loci per chromosome and sufficiently larger numbers of generations, the entire 7-chromosome genome of barley might conceivably amalgamate, sooner or later, into very few superior multilocus complexes. But this had not happened by generation $F_{46}$; disassociations also occurred, especially among loci located on different chromosomes, with the result that the numbers of differing detectable small clusters of loci continued to be large. The picture of genetic change that emerged when many different highly homozygous lines were isolated was that each individual line had improved in various ways, particularly in yielding ability and in resistance to increasing numbers of damaging races of the most serious diseases. However, it also became evident from study of many individual lines that natural selection had steadily reduced the frequency of clearly inferior genotypes and had preserved successions of superior isolates, each marked by different combinations of alleles. It was disappointing, however, that the system of multilocus markers did not identify any single spectacularly superior multilocus survivor, other than the *2112* configuration, even after 46 generations of natural selection. Yet, it was encouraging that the yield of the latest generations, en masse, was at least equal to that of recent commercial varieties and that the performance of many individual lines also appeared to be superior when compared on the basis of overall phenotypic appearance. The observed changes in single-locus allelic frequencies and in multilocus frequencies over generations were attributed to interplay between various evolutionary forces among which natural selection, acting in a temporally heterogeneous environment, was the principal guiding force. Each year selection appeared to move the population in the direction of an overall population structure most appropriate to the set of environmental factors acting in that year. Apparently, however, no single environmental regime continued for long and the direction of selection frequently changed. CCII remained highly variable genetically, and its multilocus organization appeared to fluctuate about a rarely, if ever, realized "adaptive peak" corresponding to the genetic structure appropriate to the weighted means of the environmental conditions that had prevailed during many preceding generations. Thus, the genetic structure of CCII continued to undergo successive reorganizations throughout the more than 46-generation period in which CCII was monitored. In this connection it should be noted that studies of a subpopulation of CCII that had been transferred to a site in which moisture is only rarely limiting developed multilocus associations of alleles that were quite different from those favored at the headquarters site, a site usually characterized by adequate rainfall. Conversely, in a subpopulation that had been transferred to a site characterized by moisture stress in many years, the multilocus associations of alleles favored in dry years at the central location increased in frequency. Nevertheless, the transported populations remained substantially poly-morphic during 10 or more generations in their new and environmentally more stable sites (rainfall more consistent either in deficiency or in excess) whereas the yields of the population from the headquarters site where rainfall was more variable turned out to be higher on average at all three locations than the yields of the transported populations. Evidently, the two transported populations had been less successful in moving toward the summit of a multilocus "adaptive peak" that allowed the population to do well in years of drought as well as in years of ample moisture. This result seems to point to the conclusion that environmental pertur-bations play an important role in the evolutionary dynamics of populations and that studies based on only a few generations or test environments are unlikely to provide adequate characterizations of the nature and consequences of genetic change in cultivated populations.

## WORLDWIDE PATTERNS OF MULTILOCUS
## STRUCTURE IN BARLEY

It has been recognized for many years that cultivated barleys from different eco-geographical regions of the world differ from one another in both single-locus allelic frequencies and in multilocus genotypic associations and that information about the ecological basis for such differences can be useful to breeders. Associations among alleles of different loci have conventionally been assessed in terms of two-locus disequilibrium measures. But additional complexities develop when more than two loci are considered simultaneously. This is because two major difficulties are encountered in extensions of genetic analyses that incorporate more than two loci. The first difficulty is that the number of parameters, and the number of terms included in each parameter, both increase rapidly as the numbers of loci increase, quickly leading to unmanageable complexity. Second, the structure of any given multilocus array cannot be characterized in its entirety by examining individual parameters, because the whole is not merely a summation of individual parts. Techniques for discrete multivariate analyses have now become available, and additional properties of test criteria have been characterized (Fienberg 1980). As a result, log-linear likelihood-ratio tests have become useful for analyzing complex cross-classified data. Although comprehensive descriptions are beyond the scope of the present discussion, a useful purpose may be served by outlining the major steps involved in the analysis of complex data sets such as the data set obtained by multilocus electrophoretic assays of 1,032 barley accessions chosen at random from the World Barley Collection of the USDA by Zhang et al. (1990). Fifty-two countries were represented in this sample; consequently, only the most frequent allele, plus a second composite, seemingly "sympathetic" allele compounded of all remaining alleles of the same locus, were analyzed to reduce the unmanageable multilocus situation to diallelism. Three major steps were involved in such "diallelic" analyses: (1) identification and elimination of terms that had inconsequential effects in multilocus associations, (2) construction of the log-linear model that best described the diallelic multilocus structure of the reduced genetic system under analysis, and (3) evaluation of the level of significance of each of the terms included in the model. The results showed that multilocus genetic structure differed significantly in various ecogeographical regions of the world. The genetic structures of barleys of Southwest Asia, the center of origin of cultivated barleys, are much different from those of the barleys of Europe and those of the barleys of the Far East. The barleys of the North American continent were also found to differ in multilocus structure in all of the major ecogeographical regions of North America. Cultivated North American barleys are derived from recent introductions from various parts of the world. The random sample analyzed by Zhang et al. (1990), therefore, almost certainly included varieties from many different ecogeographical regions of the world and thus included varieties adapted to habitats similar to some in North America. It is thus not surprising that the total North American sample studied had no obvious structure of its own. Overall, the individual patterns observed provide support for long-continued selection on a multilocus basis as the single most important cause of genetic organizational differences between different ecogeographical regions, as well as between subgeographical regions within larger areas.

## EVOLUTION OF ADAPTEDNESS IN
## CULTIVATED *AVENA SATIVA* IN SPAIN

The evolution of adaptedness in cultivated self-pollinating hexaploid *Avena sativa* ($n = 21$ chromosomes) is closely parallel to that of self-pollinating wild tetraploid ($n = 14$ chromosomes) *A. barbata* (described in Chapter 10). Log-linear analyses of a number of allozyme loci of *A. barbata* have established that rainfall and temperature are both closely associated with specific allelic combinations (Garcia et al. 1989). But the diverse and patchy environments of Spain lead to more complicated patterns of diversity than that in California (Allard et al. 1993). In particular, a cold-tolerant, 4-locus genotype (*Pgdl 1133, Lap1 2233, Prx1 1111, and Estl 5577*) is found exclusively within an area of Spain in which there are more than 80 days of frost per year. Correlations have also been observed in Spanish winter landraces of cultivated oats (*A. saliva*) by Perez de la Vega et al. (1994): two alternative allelic multilocus associations involving 6 loci were correlated within the warmer areas (designated the Badajoz association) versus the colder areas (designated the Soria association) of Spain. Multivariate log-linear analyses showed that allelic combinations at 4 allozyme loci were closely associated among themselves and with other factors of the environment. However, the 4-locus Badajoz association is easily extended to a 6-locus association by adding loci *Prx1 and Adh1*: more than 93% of the 4-locus plants of these five provinces also have the locus allelic association, whereas only 39% of the plants of the 4-locus Soria association have the particular Badajoz 6-locus association. Thus, although allelic combinations at 4 loci are responsible for adaptation to warm temperature regions in Spain, adaptation to colder habitats involves different alleles at the two additional loci, *Prxl* and *Adh*.

Data on other *Avena* species further implicate temperature as the environmental factor of major significance governing their allelic distributions in Spain. In contrast, the most decisive factor in California is rainfall; rainfall and its temporal distribution are clearly not such limiting factors in Spain. There have been a number of examples of associations between adaptive properties and alleles and/or multilocus combinations of alleles in other species; however, the isozyme and DNA markers in *Avena* are under the simplest possible genetic control, thus providing apparently straightforward approaches to understanding genetic aspects of the evolution of adaptedness in several species of *Avena*. We now turn to use of simply inherited molecular markers in attempting to understand long-term, as well as recent rapid and well-documented, advances that have been made in improving the yields of corn.

## ALLELIC AND GENOTYPIC DIVERSITY IN CORN

It is widely accepted that modern North American corns were derived by introgression of alleles from early cultivated Mexican dent corns into eastern North American races of flint corn to form open-pollinated varieties. Table 11-4 shows changes in the number of alleles that occurred in 23 allozyme loci during a period that extended over many centuries in the Americas. In total, 226 alleles have been found at these 23 allozyme loci in present-day teosintes and corns from many places in the Americas (an average of 9.83 alleles per locus); there is no way of knowing how many alleles were lost during the earlier histories of teosinte and corn. However, allelic diversity decreased from 226 to 163 in present-day Mexican landraces. Only 2 among the original 97 alleles present in frequency < 0.01 survived into the set of 30 popular, most popular U.S. inbreds, and only 1 of these 97 alleles survived into the set of 6 elite inbreds (Table 11-4). However, all 18 of the alleles present in high frequency (0.78–0.99) in the Mexican landrace varieties survived into the set of 30 inbreds, and all 18 also survived into the set of 6 elite North American inbreds (Table 11-4). The remaining alleles were also present in modern-day landraces of Mexico, but in low-to-intermediate frequencies. Moreover, all 10 alleles present in relatively high frequencies of 0.21 to 0.68 and 10 of the 25 alleles present in modest frequencies (0.05–0.19) survived into the 30 most popular inbreds; however, only about half of these alleles survived into the set of 6 most widely used inbred lines, and only 1 was fixed in the 6 elite inbred lines. Thus, in corn, as in oats and barley, frequent alleles contributed to adaptedness in many habitats and survived many cycles of 2 selection in cultivation, whereas alleles that were present in lower frequencies overall turned out to be useful in some, but not in other, environments; rare alleles of marker loci appear to mark chromosome segments that are of little value anywhere.

The data of Table 11-5 show that substantial reorganization occurred at the single-locus level during the breeding of the elite public single-cross hybrids and that the reorganizations that occurred at single loci ran nearly entirely counter to expectations based on the widely accepted hypothesis that heterosis is due to the advantage of homozygotes over heterozygotes at single loci. Under the overdominance hypothesis, the expectation is that the parental inbreds of high-performing single-cross hybrids will have diverged maximally in allelic frequencies so that most loci will be heterozygous in single-cross combinations. However, among the 23 loci

TABLE 11-4   Frequency of alleles of 23 allozyme loci in 94 Mexican landrace collections of maize, in the 30 most popular inbreds and in the 6 inbreds most widely used in elite single crosses (adapted from Doebley et al. 1985)

| ALLELIC FREQUENCY IN MEXICAN RACES | % OF MEXICAN COLLECTIONS IN WHICH OBSERVED | NO. OF ALLELES IN FREQUENCY CLASSES | NO. SURVIVING ALLELES INTO: | |
|---|---|---|---|---|
| | | | SET OF 30 INBREDS | SET OF 6 INBREDS |
| 0.99–0.78 | 100% | 18 | 18 (100%) | 18 (100%) |
| 0.68–0.21 | 85% | 10 | 10 (100%) | 6 (60%) |
| 0.19–0.05 | 36% | 13 | 10 (77%) | 5 (38%) |
| 0.05–0.01 | 12% | 25 | 9 (36%) | 4 (16%) |
| < 0.01 | 2% | 97 | 2 (1%) | 1 (1%) |

[a]Reid heterotic group; B73, B37, A632; Lancaster heterotic group: C103, Oh43, Mo17: these six elite inbreds are presently the most widely used inbreds worldwide.

TABLE 11-5   Twenty-three locus allozyme genotypes in $F_1$ single-crosses among the six most widely used public inbred lines in maize in North America[a] (adapted from Goodman and Stuber 1980)

| FOURTEEN[b] MONO- MORPHIC LOCI | NINE POLYMORPHIC LOCI[c] | | | | | | | | | TWENTY-THRE E LOCUS GENOTYPES |
|---|---|---|---|---|---|---|---|---|---|---|
| | 1 | 2 | 3 | 4 | 5 | 6 | 7 | 8 | 9 | |
| 11 | 12 | 11 | 11 | 12 | 12 | 11 | 11 | 11 | 11 | I |
| 11 | 12 | 11 | 11 | 11 | 11 | 12 | 12 | 11 | 33 | II |
| 11 | 12 | 12 | 12 | 11 | 11 | 11 | 12 | 12 | 13 | III |
| 11 | 12 | 12 | 11 | 11 | 11 | 12 | 12 | 12 | 12 | IV |

[a]Reid heterotic group: B73, B37, A632; Lancaster hetroic group, C103, Oh43, Mo17.
[b]*Adh1, Cat3, Ep, Got1, Got2, Got3, Idh1, Mdh1, Mdh2, Mdh4, Mdh5, Me, Mmm, Pgml.*
[c]*1 Glul, 2 Idh2, 3 Pgdl, 4 Pgdl, 5 Pgd2, 6 Pgm2, 7 Phil, 8 Est8, 9 Acpl.*

of Table 11-5, maximum divergence occurred for only 1 locus, *Glu1*; all three inbreds of the Reid heterotic group carry allele *1*, whereas all 3 inbreds of the Lancaster group carry allele *2* so that the Reid × Lancaster single-cross hybrids are all *12* heterozygotes for locus *Glu1*. Contrariwise, both the Reid and Lancaster inbreds carry the same allele for 14 of the 23 loci (61%), so that all Reid × Lancaster single crosses are *11* homozygotes for these 14 loci. The 6 most widely used inbreds were sometimes fixed for different alleles of the 8 remaining loci; consequently, some Reid × Lancaster single crosses were homozygous and some were heterozygous for these 8 loci. Overall, the data of Table 11-5 show that 79% of loci were homozygous and only 21% were heterozygous in the most widely grown public single crosses. This high proportion of homozygotes among the 23 loci of the 4 most widely grown single-cross hybrids casts doubt on the importance of single-locus heterozygosity and associated overdominance in promoting high performance.

It is widely recognized by farmers and breeders that saving seed produced on $F_1$ plants results in yields that are about 10–30% lower than the yields on the single-cross hybrids themselves; consequently, farmers routinely purchase $F_1$ seed each year. All plants in an $F_1$ population are genetically identical, so that saving seed produced on open-pollinated $F_1$ plants is equivalent to a single generation of selfing. It is also widely recognized that even a single generation of selfing such highly heterozygous materials leads to dramatic destruction of the favorable population structure of the $F_1$ generation, almost certainly because segregation and recombination, even in a single generation, breaks up the very large numbers of favorable pairwise 3-locus, up to 23-locus, interactions among the 23 loci built up by a very large number of generations of selection during the breeding of the $F_1$ hybrids. Once such favorable multiallelic interactions have been developed, it is important that such interactions be preserved and enhanced. The most effective way to preserve and enhance favorable combinations of alleles is to cross elite materials with elite relatives. Furthermore, close inbreeding of such crosses is preferable, because the wider the cross, the greater the chances that segregation will dismantle the favorable epistatic combinations; the difficulties of avoiding such disruptions depend on the numbers and alleles to be introgressed into the elite materials. Ways of doing this efficiently will be discussed in detail in Chapter 14.

## Implications for Plant Breeders

The main guiding principles for breeders that emerge from the aforementioned studies can be summarized in three concepts. The first is that the most useful genetic resources in corn are the parents of modern elite single crosses that are well adapted in local environments. Natural selection, in combination with breeder-directed selection practiced over numerous generations, has increased the frequencies of favorable alleles in the parents of such single crosses and, more particularly, the frequencies of favorable epistatic combinations of alleles, while decreasing the frequencies of less favorable alleles and the less favorable multiallelic combinations. But having many favorable alleles and favorable epistatic combinations of alleles present is not enough. Higher-order multilocus combinations of alleles, and particularly weblike combinations of alleles of different loci (Wright 1968) that tie much of the population genotype together, are also important. This leads to a second concept, namely, that once favorable multiallelic combinations have been developed for a given habitat, it is important that such combinations be preserved, added to, and expanded to include additional favorably interacting loci. The most effective way to preserve favorable combinations is to hybridize elite materials with closely related elite relatives. As mentioned earlier, inbreeding or backcrossing in such hybrids is preferable to wider crosses, because the wider the cross, the greater the chances that segregation and recombination will dismantle previously developed favorable multilocus combinations. However, sooner or later the need will arise to introgress favorable alleles from nonrelated materials into such systems. The difficulties of doing this clearly depend on the numbers and the heritabilities of the alleles to be introgressed. The third concept is that discretely inherited marker alleles have increasingly provided breeders with effective means of identifying, tracking, and incorporating additional short regions of chromosomes with favorable effects into elite materials. These situations can be illustrated by recent results demonstrating the three concepts proposed here.

The first, and perhaps the most straightforward, example involves a photoperiod-sensitive genic male-sterile (PSGMS) rice plant that was found as a spontaneous mutant in a Chinese cultivar in 1973. Early studies indicated that this mutant could be used to propagate itself under short-day conditions and to produce $F_1$ hybrid seeds by interplanting the mutant with normal fertile lines under long-day conditions. Thus PSGMS rice appeared to offer opportunities for replacing the widely used "three-line" (male-sterile, maintainer, restorer) system with a more efficient "two-line" system. However, this turned out to be more difficult than originally supposed, because it was found that in addition to photoperiod, temperature also played a significant role in the fertility of the PSGMS rice. This problem was solved by marker-assisted dissection of fertility; it turned out that nearly all of the PSGMS effect, including the temperature effect, was governed by two genetic loci, one very tightly linked to two restriction-fragment-length polymorphisms (RFLPs), one on chromosome 7 and another on chromosome 3 of rice (Zhang et al. 1995). These RFLPs made it possible to track the PSGMS effects so precisely that it was possible to avert laborious, costly, and time-consuming measurements of fertility.

The two earliest hypotheses advanced in explanation of the genetic basis of heterosis were both based on single-locus theory. Fasoulas and Allard (1962), using morphological variants marking pairs of loci situated within very small chromosome segments of barley, obtained evidence that two-locus epistasis played a significant role in the expression of 7 among 8 quantitative characters in pairs of highly isogenic lines of cultivated barley. Wright (1968), in a notable display of intuitive reasoning, postulated extensive interactions among multiple loci such that each allelic substitution would be expected to play a role in the inheritance of many traits, especially quantitative traits governed by many loci. However, multilocus assessments of these hypotheses regarding epistasis did not become feasible until large numbers of discretely recognizable Mendelian markers, such as allozymes and RFLPs, became available. When multilocus analyses became feasible, they soon provided evidence that epistasis, acting to assemble and maintain favorable multilocus genotypes, is indeed a major mechanism (perhaps *the* major mechanism) responsible for adaptedness in populations of various plant species, especially in natural plant populations. Quite surprisingly, however, little evidence for epistasis was found in a number of molecular-marker-based studies of yield and yield-associated traits in cultivated plants. Yu et al. (1997) have recently provided such evidence in an expertly conceived and carefully executed experiment designed to characterize the genetic basis of heterosis at the single-locus and two-locus levels in a remarkably heterotic rice hybrid, Shanyou 63. Their experiment, in addition, provided evidence for multilocus epistasis. Shanyou 63, the best rice hybrid of China, has been grown on 6.7 million hectares annually during the past decade, accounting for approximately 25% of the very large rice production of China. This study serves as a second example of marker-assisted dissection of adaptedness in cultivation.

The experimental population of Yu et al. (1997) was made up of 250 $F_3$ families, each descended from bagged panicles descended from a single $F_2$ plant of the hybrid between the two parents (Zhenshan 97, Minghui) of Shanyou 63. The $F_3$ families, the two parents, and the $F_1$ and $F_2$ generations were transplanted to a bird-net–equipped field in an experimental area of the Huazhong Agricultural University, Wuhan, China, in the 1994 and 1995 rice-growing seasons. The field plantings were arranged in randomized complete block designs with three replications in both test years. Within each replication 35-day-old seedlings were transplanted in a single-row plot, at a distance of 17 cm between plants, within rows 17 cm apart. Field management thereafter closely followed normal agricultural practice as closely as possible. Each plant was harvested individually at maturity to prevent seed loss from overripening. Traits examined included yield per plant (converted to metric tons/hectare), the number of seed-setting tillers per plant, grain size (weight in g/1,000 seeds), and grains per panicle. Two types of DNA markers, RFLPs and simple sequence repeats, were used to assay polymorphisms in the parents, $F_1$, $F_2$, and the $F_3$ generations. The entire genome was searched for digenic interactions in the $F_2$ and $F_3$ generations for each trait by means of two-way analyses of variance using all possible two-locus combinations of marker genotypes. The 8 degrees of freedom among the 9 genotypes formed by each of the two codominant loci included 2 degrees of freedom for the additive and dominance effects within each locus and 4 degrees of freedom for the interactions among loci. The latter interactions, representing epistasis, were further partitioned into four terms, each specified by a single degree of freedom as follows: additive (first locus) × additive (second locus) that is, A × A (AA); additive × dominance (AD); dominance × additive (DA) and dominance × dominance (DD). The statistical significance of each term was tested using an orthogonal contrast test. $F_1$ yields, as well as the three yield-component traits were approximately the same in the two test years. Heterosis, measured as the percentage deviation from mean parental yields, was large in both years, and residual heterosis in the $F_3$ generation was also large, whether measured in terms of seed yield, tillers per plant, grains per panicle, or grain size. A survey of 537 RFLPs and 54 simple sequence repeats identified 150 differences between the two parents that could be assigned to 14 linkage groups. A total of 32 distinct quantitative trait loci (QTLs) were identified, among which 12 were observed in both years and the remaining 29 were detected in only the first or the second year. Several QTLs appeared to have pleiotropic effects; that is, they affected two or more traits simultaneously.

Correlations between overall heterozygosity of $F_2$ individuals and $F_3$ family means were very small for all four traits measured. Thus, overall, heterozygosity apparently made very little contribution to trait expression. However, various levels of negative dominance were observed in both years for many QTLs, which suggested that cancellations between positive and negative dominance might have been responsible for the small overall overdominance that had been observed for many traits in previous studies. As emphasized by the authors, perhaps the most important finding of this study was the prevalence and the importance of epistasis in the rice genome, and, more particularly, that the epistasis had two pronounced features. First, the two-locus analyses detected far larger numbers of loci contributing to trait expression than the single-locus analyses. For example, for the trait grain number per panicle, the numbers of loci involved in two-locus trait expressions included a total of 25 loci located on 9 of the 12 rice chromosomes in the two years combined, as compared with only 5 single-locus cases in first year and only 7 in the second year. A second feature was that all four types of interactions, AA, AD, DA, and DD, were found among the 25 two-locus combinations. Even more remarkable was that multiple interaction terms were found in a very high proportion of the two-locus combinations. Furthermore, the results indicated the likelihood of higher-order interactions (multilocus epistasis), especially for the most complex trait (yield). Two lines of evidence suggested the existence of higher-order interactions among different loci. First, far fewer QTLs were found, and a smaller proportion of the genotypic variation was explained by QTLs for yield than for the yield-component traits, although the reverse would be expected on the basis of multiplicative relationships of yield with component traits. In addition, at the two-locus level the numbers of interactions detected for yield were not much greater than those detected for the yield component traits. This indicates the involvement of genetic components that could not be resolved with either single-locus or two-locus analyses. This idea became still more plausible when significant two-locus interactions revealed "weblike" relations among the interacting two-locus combinations such that locus *1* interacted with locus *2*, which in turn interacted with locus *3*, locus *3* with locus *4*, and so on. Such webs could grow very large, as previously postulated by Wright (1968) and as had been observed in an experimental population of barley (Allard 1990) as well as in natural populations of *Avena barbata* (Perez de la Vega et al. 1991), in cultivated hexaploid *Avena sativa* (Perez de la Vega 1996), and in cultivated populations of corn.

In summary, the results of marker-assisted dissections of quantitative traits have clearly established that epistasis plays a major role in the inheritance of quantitative traits as well as in the genetic basis of heterosis. The relationships between traits and genes in the manifestation of heterosis are much more complex than expected on the basis of either the dominance or overdominance hypothesis. Digenic interactions, including additive × dominance and dominance × dominance interactions, were frequent and widely dispersed throughout the genome in all of the aforementioned populations. The interactions involved large numbers of different marker loci, most of which were not detectable on a single-locus basis. These results provide definitive evidence that epistasis plays a major role, and probably even the major role, in the genetic basis of heterosis.

# PART III

*Modern Breeding Plans*

# TWELVE

# *Reproductive Systems and Breeding Plans*

Modern plant-breeding plans differ from crop to crop in various ways, depending on the purposes for which different crops are grown and on their various attributes. Nevertheless, many otherwise apparently quite unlike crops have particular attributes in common, especially mating system and/or method of propagation, such that nearly all crops can be assigned explicitly to one of three main groups in respect to the breeding plan that will be appropriate for each:

*Group 1*—primarily self-fertilizing, seed-propagated annuals such as wheat, rice, barley, and Phaseolus beans (see Table 2-1)
*Group 2*—outcrossing, seed-propagated annuals (especially corn), biennials, short-lived perennials, and long-lived perennials, including many timber tree species
*Group 3*—vegetatively (clonally) propagable outcrossing perennials, many of which can also be propagated from seeds

The population structures of the wild progenitors of these three groups of plants, as well as the population structures that have developed in cultivation, are closely similar within each of the three groups. All Group 1 species have remained highly tolerant of continued selfing in cultivation, and these selfing species, like their wild ancestors, typically are made up of mixtures of usually rather limited numbers of highly homozygous, nonsegregating, but differing, multilocus genotypes that are highly adapted to the local environment. Outcrosses occur occasionally between individuals with different multilocus genotypes within any given population. When such intrapopulational $F_1$ hybrids reproduce by selfing, as is usually the case (98% or more of the time), the segregation and recombination that occurs in the first few generations following leads to highly rigorous linebreeding that, in combination with selection, tends to concentrate favorable combinations of alleles into novel homozygous multilocus genotypes, among which some are likely to be superior to the parental genotypes from which such segregant hybrids were formed. Occasionally $F_1$ hybrid plants (or later-generation offspring of $F_1$ hybrids) hybridize with a genetically quite different plant in the same population. This also leads to highly disciplined linebreeding that, within a few generations, is likely to introduce additional homozygous (true-breeding) genotypes into the population. The advantage of this system of mating and reproduction is that it provides automatically for orderly, ongoing, and continuing progressive insertion of superior genotypes into the population and, hence, almost automatically leads to improvement in performance.

Group 2 species, in contrast, engage in extensive outcrossing that typically occurs among the large numbers of genetically very different individuals that are likely to be present in all generations. Thus, all individuals in Group 2 species result from wholesale indiscriminate outcrossing among many genetically different plants, including new hybrids often untested or only briefly tested for adaptedness to the local environment. Development of population structure based on superior multilocus assemblages of genotypes capable of providing close and high adaptedness to the local environment in Group 2 plants is plainly much less disciplined and much less orderly than is the case with the nearly ubiquitous selfing that leads to directed within-line breeding that occurs perforce in heavy selfers. Nevertheless, in Group 2 species alleles that have individually favorable effects on adaptedness are expected to increase in frequency over long periods of time and, sometimes, to amalgamate with other favorable alleles to form novel favorable multilocus genotypes. This process, too, will almost certainly be much less orderly and much less effective than is the case with Group 1 species. This may well be the reason that yields increased so very slowly during the many years that corn was selected in open-pollinated random mating populations (Figure 5-1). Evidence that cross-fertilization in outcrossing organisms can be beneficial had been reported by Vilmorin (1856), Darwin (1859, 1876), and others, but the critical observation came in 1908 when G. H. Shull concluded that "the correct field practice in breeding corn must have as its object the maintenance of such hybrid combinations as prove most vigorous and productive, and give all desired qualities of ear and grain." Shull had noticed that inbred lines as a group were in all cases inferior to hybrids between inbred lines. These observations forced him to a previously elusive and unappreciated point, namely, that it was crossing that had advantageous effects in outcrossing species and that the breeding of corn must have as its objective identification of those particular hybrid combinations that are most vigorous and productive. Shull's concepts adumbrated the intricate and complexly interrelated procedures used today in breeding hybrid varieties of corn and other outcrossing crop species, as well as in the breeding of animals. Jones (1917) thereafter soon identified methods for capitalizing on favorable interactions among exceptional inbred lines. The great additional progress that has subsequently been made in capitalizing on favorable interactions among inbred lines of corn by enhancing the value of already good inbreds will be discussed in detail in Chapter 14.

Many clonally reproducible Group 3 species, notably strawberries and potatoes, are capable of producing large numbers of seeds; such species are typically bred by crossing desirable clones and examining the large numbers of genetically different genotypes that result from segregation and recombination following such sexual reproduction. Note, however, that in vegetative reproduction, vegetatively produced progeny of each clone will be uniform and will continue to exploit all types of genetic advantage (including heterozygote advantage and favorable epistatic interactions) in generation after generation of vegetative propagation. Hence, once superior types have been identified in sexually produced progenies, such types can be reproduced clonally from tubers, cuttings, stolons, or other vegetative propagules and, in some cases, treated thereafter agriculturally as annuals, as has been done with strawberries and potatoes. However, some agricultural populations of clonally reproducing species (e.g., populations of sugar cane) are grown commercially as complex mixtures of clones developed by selection among seedling segregants from hybrids between superior parental clones. All vegetative progeny of such segregants, although likely to be highly heterozygous, are thereafter genetically identical, owing to descent by vegetative reproduction. Usually, selection among clonal segregants is practiced in cultivation in such complex mixtures with the goal of eliminating inferior clones and at the same time increasing the frequency of superior clones in mixed populations. Breeding clonally propagated crops will be discussed in detail in Chapter 16.

## GROUP 1 CROPS

The largest and by far most important group of food crops are Group 1 species, which include several highly self-pollinating cereal grasses and feed grains, such as the wheats, rice, barley, sorghum, millets, a large number of cereal legumes (e.g., soybeans, common and lima beans, peanuts, lentils), and various vegetables such as tomatoes and peas. Populations of the wild progenitors of such crops, as well as the crops themselves, are sometimes made up of only one homozygous true-breeding line or of mixtures of closely similar highly homozygous true-breeding lines. Outcrossing is usually infrequent, so that outstanding homozygous pure lines often persist in each such population for large numbers of generations. Many of the natural outcrosses that occur over time in such populations are between genetically identical plants and, hence, like self-pollinations, they do not change the effect on the genetic composition of the population. However, occasional outcrosses are between genetically different individuals, and when such outcrossed individuals reproduce by selfing, as is almost always the case, the effect of the selfing is rigorous "linebreeding" owing to the heavy selfing. During the first three to five generations of selfing of such hybrids, homozygosity decreases by half per generation so that by generation $F_4$ or $F_5$ nearly complete homozygosity is restored in most field families, and some among the many novel homozygotes that appear, particularly if they are adaptively superior, are likely to be incorporated into the population. Occasionally, novel segregants hybridize with other plants in the same population, initiating additional rounds of linebreeding that, within a few generations, often introduce still further superior homozygous genotypes into the population. The advantage of this mating system is that it provides, automatically, for orderly and progressive insertion of many superior true-breeding genotypes into each population.

## GROUP 2 CROPS

What happens in Group 2 species is sharply different: outcrosses occur almost without constraint among many different individuals in all generations. Thus, most individuals in Group 2 species result from random, indiscriminate outcrosses among many genetically different plants in each generation. The inevitable result is that, generation after generation, all populations of such species are made up of segregating and constantly changing arrays of differing alleles at many loci and, more particularly, constantly changing arrays of differing multilocus genotypes that are likely to be eliminated before they can be tested for adaptedness. Development of population structure based on adequately tested multilocus assemblages of genotypes capable of providing close and high adaptedness to the local environment in Group 2 plants is thus plainly highly undisciplined and clearly much less orderly than is the case with the ubiquitous inwardly directed linebreeding that results from the almost automatic selfing that occurs in heavily selfing populations. Nevertheless, over long periods of time, alleles that have individually favorable effects may be expected to increase in frequency and gradually amalgamate into favorable multilocus genotypes; however, amalgamations favorable to long-term survival must be largely serendipitous events in outcrossing populations. This is almost certainly the reason that corn yields increased so very slowly during the many years corn was grown and selected in open-pollinated populations (Figure 5-1). When breeder-imposed selfing, which produced large numbers of ultimately true-breeding homozygous inbred lines, became commonplace in corn, from the late 1920s through the 1940s, this change in breeding system provided for quick formation of favorable combinations of alleles resulting from selfing and for survival to homozygosity of many favorably interacting multilocus combinations of alleles as inbred lines were selected. Fortunately, the notion that cross-fertilization in outbreeding organisms can be beneficial (Vilmorin 1856, Darwin 1859, 1876, Shull 1907, Jones 1917) was recognized, and methods for identifying those rare inbreds that complement one another genetically to produce genetically superior hybrids improved rapidly in the following decades. Methods for subsequent continuing improvements of inbreds to capitalize on still more favorable interactions among lines will be considered in detail in Chapter 14.

## GROUP 3 CROPS

Group 3 species are typically composed of very complex mixtures of different vegetatively reproducing clones, nearly all of which are homozygous at some loci but heterozygous at many other loci. In wild Group 3 species allelic frequencies, as well as multilocus genotypic frequencies, are likely to stabilize ultimately at levels that provide for elevated but often less than optimum fitness. Among species that are capable of vegetative (clonal) reproduction, the many vegetatively reproduced lines that can be produced from each clone will be identical (except for rare somatic mutations); hence, each such clone is potentially immortal. There is evidence that some clones in nature are hundreds of years old and that occasional natural populations have become uniform for a single evidently superior clone. This suggests that a clone that fits very well into a particular existing environment may have arisen long ago and ultimately come to dominate additional sites with the same or a similar environment.

The practice of clonal propagation is truly ancient in agriculture, and some present-day clones appear to be identical to clones that have been cultivated for thousands of years. Perhaps the main thing recent intervention by humans has accomplished for clonal crops is the substitution of deliberate selection within sexually produced seedling progenies for selection within occasional natural inter-clonal hybridizations. This has been particularly effective, especially recently, with species such as strawberries and potatoes that produce large numbers of seeds sexually but have the additional advantage that each sexually produced individual can be propagated clonally and thereafter treated as an annual. Note that such vegetatively reproduced genotypes remain identical, vegetative generation after generation (in the absence of mutations), when reproduced vegetatively, so that replicates can be produced at will for the purpose of rapid evaluation of agricultural potential in performance trials replicated over time at many locations. Moreover, selection practiced in cultivation among very large numbers of clonal segregants often quickly identifies segregants that are superior in many environments. Some agricultural populations of vegetatively reproducing species (e.g., sugar cane) are grown commercially as complex mixtures of clones, which are subsequently rese-lected to reduce the frequency of inferior clones and increase the frequency of superior clones in the mix.

Some present-day clones (e.g., clones of date palms) appear to be identical to highly satisfactory clones known to have arisen in very ancient agricultures. It has been suggested that developing replacements for such slow-growing materials would be a lengthy process and unlikely to be cost-effective. More detailed discussion of "modern" breeding methods that have been successful with clonally propagable species is deferred until Chapter 16.

It was not until early in the twentieth century, more than 10,000 years after plants were first taken into cultivation, that the mating patterns most appropriate to various specific breeding goals were clearly defined by Wright (1921). These mating patterns are (1) random mating, (2) genetic assortative mating, (3) pheno-typic assortative mating, (4) genetic disassortative mating, and (5) phenotypic disassortative mating. In plant breeding the individuals chosen to produce the succeeding generations, as we saw in Chapter 5, have until recently been allowed to mate at random by haphazard chance, as was the case in nature before crop plants were domesticated. This leads to very different population structures in outbreeders as contrasted with predominant selfers, whether in nature or in cultivation. Wright's four other sexual patterns of reproduction, in addition to random mating, all represent deviations from random mating based either on deliberate like-to-like matings or on deliberate matings of unlike individuals. The criterion of likeness or unlikeness is either relationship (ancestry) or appearance (phenotypic resemblance vs. phenotypic difference). A sixth common reproductive scheme in plants is vegetative reproduction (clonal reproduction). Vegetative reproduction, of course, immediately causes the genetic relationship between parent and offspring to become complete; that is, the genetic parent-offspring regression immediately becomes 1.00, and it remains so generation after generation in the absence of spontaneous mutation.

With genetic assortative mating, the mating individuals are more closely related by ancestry (descent from a common ancestor) than if the matings had been at random. With phenotypic assortative mating, the mating individuals resemble each other more closely in phenotypic appearance than randomly chosen individuals. The third nonrandom system, genetic disassortative mating, calls for matings between individuals that are less closely related than randomly chosen individuals. Finally, with the fourth nonrandom system, mates are chosen on the basis of unlike phenotypic appearance; this constitutes phenotypic disassortative mating. Thus two processes, (1) choice of parent or parents and (2) choice of mating pattern, are jointly the primary operations by which breeders guide changes in allelic and genotypic frequencies in desired directions in their breeding populations.

The primary postulate of population genetics is that allelic frequencies and the distributions of alleles among individuals in any filial generation are determined not only by the selective values of the alleles inherited from the parents but also by the mating system in force. Thus, for example, under truly random mating the zygotic proportions for selectively neutral alleles $a_1$ and $a_2$ at any diallelic locus ($p_2\,a_1\,a_1 : 2_{pq}\,a_1\,a_2 : q_2\,a_2\,a_2$) no longer pertain under any other system of mating. In the discussions to follow we will examine how changes in zygotic proportions that result from applying various mating systems influence the extent of genetic variability, levels of heterozygosity, and genetic correlations among relatives, as well as the genetic structures of populations.

## RANDOM MATING

Strictly speaking, the term *random mating* should be applied only to situations that meet two criteria: (1) each member of a population has an equal chance to produce offspring and (2) any (female) gamete is equally likely to be fertilized by any (male) gamete. It can be questioned whether the theoretical form of random mating is ever exactly fulfilled in plant breeding, because some form of selection, natural or human-imposed, is likely to intervene at some stage or stages in the life cycle, thus violating the first criterion (equal chance to produce offspring) and permitting the breeder to inadvertently and unknowingly favor particular plants. If this is the case, only the second requirement is fulfilled, so that the mating system is more appropriately called *random mating with selection*. Note that random mating, with selection is often likely to be closely similar to phenotypic assortative mating (the third nonrandom mating system), in which phenotypically similar individuals are identified according to some defined set of criteria and then isolated together or otherwise deliberately mated together. This differs from pure random mating, in which the choices of the individuals to be mated are assumed to be entirely chance events. It should further be noted that there frequently is also doubt as to whether the second criterion is ever fulfilled strictly in plant breeding. Plant-to-plant differences in ovule maturity, time of pollen shed, location in nurseries in respect to prevailing wind direction, and the like make it improbable that fertilizations are ever entirely at random. If, however, all conditions necessary for the theoretical form are met, three predictions can be made concerning the genetic composition of large outbreeding populations:

1. Allelic frequencies will remain constant from generation to generation.
2. Levels of heterozygosity and homozygosity will remain constant, in accord with Hardy-Weinberg proportions.

3.  Genetic relationships among individuals will also remain constant over genera-
    tions at a value of $F = 0$ of the inbreeding coefficient.

The situation is expected to be the same in large selfing populations in regard
to the first prediction; thus, in the absence of selection, allelic frequencies are
expected to remain constant in the population as a whole. However, outcomes are
expected to be very different regarding the second and third predictions. This is
because complete selfing causes any genetically heterogeneous population made up
of different genotypes, whether homozygotes or heterozygotes, to become divided
into a number of reproductively independent families, each descended from a
single founder if selfing is complete. Assuming that total population size, summed
over all families, remains constant, each originally heterozygous family is expected
to include fewer heterozygotes as segregation and recombination produce more
and more fully homozygous and genetically distinctive families. Thus, in the
absence of selection, and assuming that all families are equal in reproductive
capacity, allelic frequencies are expected to remain constant overall in large completely selfing populations as family sizes become increasingly smaller. Expectations
are, however, very different regarding the overall structure of large completely
selfing populations contrasted with large random-mating populations. The differences in overall population structure arise in the differing effects of random mating
versus selfing on predictions 2 and 3. Complete selfing rapidly drives heterozygosity
to 0 and homozygosity toward 1.00 at all loci, so that genetic relationships among
individuals soon reach and thereafter remain at $F = 1.00$ over generations. Thus a
completely selfing population, instead of remaining a homogeneous whole, sepa-
rates within a few generations into numerous genetically distinct but internally
homogeneous (homozygous) coexisting families, with the result that the size of any
single such family within any selfing population is likely to be small unless total
overall population size increases. Note that families that happen to receive more
than the average number of favorable alleles are likely to benefit in fecundity from
the automatic linebreeding that occurs and, thereby, increase in frequency relative
to less favorably endowed families. Note also that if selfing is partial in some families,
patterns intermediate between those of complete selfing and random mating will
develop.

## Effects of Random Mating with Selection

When effective selection is practiced and the chosen individuals are mated at
random, the outcome is different from that expected with the theoretical form of
random mating. Effective directional selection reduces variation within each group
of individuals selected to be parents. Lush (1945) gave the following examples of
the magnitude of this reduction on the filial generations in random mating populations. If the breeder eliminates the 10% most extreme individuals from one tail
of a normally distributed population and retains the remaining 90% of individuals
as parents, the expectation is that the standard deviation of the retained group will
be reduced by 16%; eliminating 20% of extreme individuals from one tail reduces
the variation of the retained individuals by 24%, and eliminating 50% reduces the

variability of the retained group by 40%. Thus effective selection is expected to increase noticeably the uniformity of the group of individuals selected to be parents. The key question is, of course, what effect will such selection have on the next generation? When the selected and now more uniform parents are intermated at random, the extent of variation in their offspring will, according to Lush (1945), have been altered in one way or another as contrasted with an unselected group. If the values of $q$ are small—that is, if the alleles that affect the character in consideration are infrequent—the effect of increasing their frequency will be to increase variability of the next generation. If, however, values of $q$ are generally large, the effect will be to decrease the variability of the next generation. Regardless, these effects are likely to be small for quantitative characters, particularly for characters of low heritability. In conclusion, directional selection is not likely to have much effect on most performance characters. Patterns of change will clearly be intermediate when the population mates partly by selfing and partly by intercrossing at random.

A second effect of successful unidirectional selection will be to produce an excess of intermediate gametes by eliminating gametes produced from one or the other tail of the distribution curve. This narrowing effect is also expected to be slight; for example, if half of the individuals of a previously unselected random mating population come from the central part of the population curve, the standard deviation of the next generation will be only ~ 17% smaller than that of the original population, even in the unlikely case that heritabilities are 100% and all genetic effects are additive. But if heritabilities are only 50%, the reduction in variability will be only 4%, and if the heritabilities are only 30%, the reduction in variability in the next generation is expected to be a minuscule 2%. In summary, it is clear that the main effect of random mating with selection will be confined to changing the frequencies of simply inherited alleles that have large and highly heritable effects. With more complexly inherited characters, as well as when heritabilities are low, the ability of random mating with selection, and of mixed random mating and selfing, to increase homozygosity declines rapidly until, for most production characters, this breeding system has very poor powers in producing fixation. It was shown by Wright (1921), that with $N$ loci of equal additive effect, the standard deviation is $2(pq)^{1/2}$ times the effect of one locus and that the potential range of expression of the character is $2(pq)^{1/2}$ times its standard deviation. Individuals that deviate by 2 standard deviations or by 3 standard deviations are consequently expected only once among 21 individuals, or once among 370 individuals, respectively. The larger the number of loci, the less likely it is that the most extreme type possible will appear in random mating populations or in populations that mate by a mixture of selfing and random mating. With additional cycles of effective selection, the frequency of desired alleles is expected to increase, but only very slowly, and the result will be not only a shift in the mean of the population but the possible (and hoped for) appearance in the population of individuals with more useful phenotypes than were present in the original population.

If all of the genetic variation in a population is due to additive gene action, improvement should continue, theoretically, until all desirable alleles are fixed. However, the rate of fixation depends on both the number of loci and allelic frequencies. If only 1 diallelic locus is involved and $p = q = 0.5$, the limit of selection with random mating is $2N/_{pq} = 2.83\sigma$. If however 10 equally effective loci (all $p = q = 0.5$) are involved, the limit of selection is $8.95\sigma$, and with 20 diallelic loci (all $p = q = 0.5$) the limit of selection approaches $13\sigma$. With a single locus it is expected that $p$ (or $q$) will approach 1.0 quite rapidly, because the variance will be very small when $p$ (or $q$) is near 1.0 or 0. Even though selection may take many generations to fix a monogenic character, $\Delta q$ will be very small when $p$ (or $q$) is small; hence, selection is expected to reduce the variance of monogenic characters rapidly to near the vanishing point, giving the appearance of nearly complete exhaustion of genetic variation within a few generations. If, however, the number of loci is large, $\Delta q$ will be small for each locus, even when $p$ (or $q$) is near 0.5; consequently, $\Delta q$ will almost certainly be unacceptably small (from the standpoint of plant breeders) should allelic frequencies differ much from 0.5. As a result, only very slow progress can be expected when selection is practiced in random mating populations, even for loci that act in a largely additive fashion. Most production traits appear to be inherited in multigenic fashion; thus, mass selection with random mating can, in general, be expected to be inefficient for such traits.

It is unclear whether there are, in fact, cases of plants in which the heterozygote at any single locus is superior to both homozygotes. If truly overdominant situations do in fact exist, selection under random mating will lead to stable equilibria in allelic frequencies rather than fixation of one of the homozygotes. Representing the reproductive rates of genotypes *aa: Aa: AA* as $1 - s_a$: 1: $1 - s_A$, it can be shown that at the equilibrium point $q = 0$ is given by $s_a/S_a + s_A$; that is, the equilibrium point depends solely on the coefficients of selection of the homozygotes. Thus, the equilibrium point is independent of initial allelic frequencies in the population. It follows that fixation of type is not possible in random mating populations if the heterozygote at a locus is favored over both homozygotes. A maximum proportion of heterozygotes will occur when allelic frequencies are 1/2, but even then 1/4 of the population will continue to exhibit one or the other of the less desirable homozygous phenotypes. Thus, even though allelic frequencies of 1/2 produce the maximum proportion of the "best" phenotypic class, it does not follow that this is the best average phenotype for the population. It is at the equilibrium point ($\Delta q = 0$) that the ratio of homozygotes to heterozygotes will be such that the mean phenotypic expression of the population will be maximal. For example, if reproductive rates are $AA = 1.0$, $Aa = 1.2$, and $aa = 0.8$, maximum fitness occurs at $qA = 2/3$, at which point genotypic frequencies are 0.444 *Aa*, 0.444 *Aa*, and 0.11 *aa*, so that the mean population fitness for this locus is 1.07, a point intermediate between the fitnesses of the *AA* and *Aa* genotypes. Consequently, assuming that the relative reproductive rates of *aa:Aa:AA* are 0.0:1.2:1.0 (i.e., *aa* is sterile), zygotic frequencies at equilibrium will be 0.73 *AA*:0.25 *Aa* and 0.02 *aa*, giving a mean population fitness of 1.03. Clearly, breeding strategies other than random mating are required to maximize the usefulness of truly overdominant alleles (assuming such alleles exist). If multiple overdominant alleles exist at any locus, allelic frequencies at such loci will be proportional to the reciprocals of their respective homozygotes in random mating mass-selected populations: $q_1:q_2:q_3 = 1/s_1:1/s_2:1/s_3$. This theoretical relationship is based on the assumption that all homozygotes are, in some degree, inferior to the heterozygotes. If, however, one heterozygote (say *a1q2*) is superior to all other genotypes, alleles $a_1$ and $a_2$ will ultimately dominate the population.

## Random Mating with Nonadditive Gene Action

There are a number of possible genetic situations, in addition to selection favoring single-locus heterozygotes, that can limit progress under selection. One such situation results when interactions among different loci lead to nonlinearity on the fitness scale such that selection favors an intermediate phenotype. Suppose the frequencies of alleles A and B are both 0.5, that their effects are additive, and that genotypic frequencies in a random-mating population are as given in Table 12-1. Also suppose that a breeder has evidence that individuals with an intermediate phenotype (it has been observed that intermediate height or intermediate seed size are often favorable) are, in fact, superior to all other phenotypes and, hence, that only individuals with phenotype 3 are saved. In the first generation of selection under full random mating, this will increase the frequency of individuals with phenotype 3 from F = 6/16 (0.375) to F = 1/2 (0.50). Another generation of errorless selection will increase the frequency of phenotype 3 to F = 0.56, and still another generation will increase its frequency to about F = 0.58. This will also reduce the variance of the population for this character to about half of its original level, but beyond this point errorless selection will be almost powerless to produce further reductions in either the proportions of phenotype 3 or the variance of the population. Note also that if selection is relaxed, the population will return to its original status—that is, a mixture of individuals with phenotypes 1 to 5 (mean phenotype 3) within 3 or 4 generations. Thus, such selection does not change the frequency of either allele A or B; instead, it causes the gametic ratio to be distorted, and on relaxation of selection, recombination will soon restore the gametic ration to 1:1:1:1. Obviously, under random mating, selection must be continued indefinitely if any gain that resulted from the increased proportion of individuals with phenotype 3 is to be maintained. Note also that errors in selection (e.g., selection of individuals with phenotypes either larger or smaller than phenotype 3) will result in regression toward the original level. In this example the alleles of both loci were postulated to act in a strictly additive fashion; hence, it was nonadditivity on the desirability scale that conferred superiority on phenotype 3. There is increasing evidence that epistatic interactions among alleles of two, and often many more, loci

TABLE 12-1  Selection favoring an intermediate phenotype

| GENOTYPE | FREQUENCY | PHENOTYPE IN ARBITRARY UNITS |
|---|---|---|
| *aabb* | 1 | 1 |
| *Aabb, aaBb* | 4 | 2 |
| *Aabb, aaBB, AaBb* | 6 | 3 |
| *AABb, aaBB* | 4 | 4 |
| *AABB* | 1 | 5 |

play major roles in the development of favorable multilocus genotypes in nature and in cultivation (Chapters 10 and 11). However, multilocus combinations of alleles are likely to be transitory in random-mating populations because of segregation and, as is apparent in the preceding example, it is difficult to increase the frequency of the favorable *AAbb*, *aaBB*, and *AaBb* genotypes substantially. Furthermore, continued errorless selection is necessary to preserve the modest gains that result from selection for phenotype 3. Table 12-1 suggests a simple and straightforward solution for the breeder, namely, selecting individuals with phenotype 3 and thereafter advancing generations following a nonrandom mating scheme (genetic assortative mating = inbreeding), which, as we will see later in this chapter, is the most efficient of all mating systems for concentrating and stabilizing favorable alleles within breeding stocks.

### Factors Limiting Progress Under Random Mating

Many factors can limit progress under selection when the breeding scheme is one of random mating. We now focus on two factors that play major roles in plant breeding: (1) heritability and (2) practicable selection intensities. As a simple and often encountered situation in plant-breeding programs, consider a random-mating population in which a fully dominant and completely penetrant resistance allele $R$ is present in a population in high frequency (say $f = 0.80$), so that 96% of plants in a variety a breeder wishes to release for commercial production are resistant and 4% are susceptible. Suppose the breeder decides to remedy this untidiness by selecting resistant plants ($RR$ or $Rr$) while eliminating susceptible plants ($rr$) from the population. If no mistakes are made in classification, 12 generations are required to increase the frequency (F) of $R$ from 0.80 to 0.95 and an additional 32 generations are required to increase $R$ to F = .98. Suppose, however, that the pathogenic organism causing the disease is present only erratically, so that the coefficient of selection averaged over generations is no longer 1.0 but only .20. In this circumstance, 54 generations are required to increase $R$ from 0.90 to 0.95 and an additional 155 generations are required to increase $R$ from 0.95 to 0.98. Selection in random-mating populations is obviously an inefficient way to achieve near fixation of dominant monogenic traits, and such selection will clearly be much less efficient for multigenic characters of low heritability. These are among the reasons that breeding schemes other than random mating (especially genetic assortative mating schemes) have largely supplanted ineffective mass selection in outcrossing plant species.

A second major factor affecting progress under random mating is the intensity of selection that is possible with a given species. If all of the progeny that can be produced per generation are needed merely to perpetuate breeders' populations, it is obvious that no selection at all can be practiced. If, however, a species is very prolific, selection differentials of dozens—or even hundreds—to one may be possible, assuming that the breeder has the resources needed to produce and process such very large populations. Suppose the resources of a breeder are such that it is not practicable to grow more than 1,000 plants/generation, and experience indicates that it is necessary to save the extreme 5% of plants for the character under selection merely to perpetuate the population. In this circumstance the breeder must save the 50 most desirable plants among those present in his or her nursery each generation. It would therefore be necessary that each plant produce 20 offspring, on average, to satisfy minimal population sizes each generation. If only the extreme 1% of plants have phenotypes that qualify them as parents, the average number of offspring required to maintain population size increases to 100. For most crop species selection differentials of this magnitude do not present unsurmountable problems, but in the breeding of some large long-lived plant species (and perhaps all large animals) such population sizes are beyond possibility.

However, the problem is nearly always much more complicated because breeders are rarely able to confine attention to one character at a time. Ordinarily, it is necessary to deal simultaneously with at least two, and sometimes several, uncorrelated, weakly correlated, or negatively correlated characters. Suppose the breeder wishes to save the best 5% of individuals for both Character A and Character B; obviously, selection pressures must be relaxed for both characters if selection is to be practiced simultaneously. In this case, perhaps the best the breeder can do is to save all plants that are among the 23% best for Character A and for Character B (unless fortune smiles and some individuals are among the best for both characters). If equal attention is to be given for 3 characters and the reproductive rate requires that 5% of the total population be saved each generation, selection for each character must be confined to the best $\sqrt[3]{0.05} = 37\%$ of the individuals. With 5 characters, the breeder will be able to save only the best 55% for each character, and with 10 characters, selection must be confined to the best 74% for each character. Clearly, increasing the number of characters for which selection is practiced rapidly decreases the intensity of selection possible for any one character to very low levels, unless the breeder's resources are sufficiently large to allow huge populations to be grown and processed. Although low fecundity can be a problem in plant breeding, the high prolificacy of some crop species can lead to problems of an opposite type. Breeders of highly prolific species must be discerning in regard to the materials they save so as to avoid being overwhelmed by more materials than they have time or funds to grow and examine in future generations.

## NONRANDOM MATING: GENETIC ASSORTATIVE MATING

The primary effect of genetic assortative mating is to increase the probability that offspring will inherit favorable alleles that are alike by virtue of identity by descent from a common ancestor. Genetic assortative mating thus serves to reduce heterozygosity, increase fixation of alleles of the same locus, and, when combined with effective selection, whether natural and/or human directed, promote fixation of favorable multilocus genotypes as well as stabilize phenotypes to the extent that they are under genetic control. Genetic assortative mating, as we saw in Chapter 5, is the most effective of all mating systems for these purposes because degree of relationship owing to descent from common ancestors can be determined straightforwardly by pedigree records. Accurate assessment of relationship does not depend on ability to determine which individuals carry the same alleles or which individuals have the same multilocus genotypes, but only on ability to keep accurate records of descent. The predictive ability of genetic assortative mating is thus minimally hampered by low heritabilities, that is, failure of filial phenotypes to match ancestral phenotypes owing to environmental causes or to genetic causes such as dominance or epistasis. Genetic assortative mating is therefore particularly useful for manipulating alleles affecting traits of low heritability; these are the traits that are often most important in breeding and most difficult to manage in practical breeding.

Intensive genetic assortative mating in heterogeneous populations rapidly leads to separation of such populations into a number of noninterbreeding groups, and if overall population size is to remain within manageable limits, family sizes must be restricted or some groups must be discarded. Discarding families has the practical effect of forcing plant breeders constantly to make decisions regarding which families to discard or retain. The effect of genetic assortative mating on total genetic variability depends on the selection practiced. Inbreeding without directional selection increases genetic variability among families because, as genetic correlations among relatives within lineages increase, homozygosity increases and genetic differences among families also increase owing to random increases or decreases of different alleles within different families. However, if effective directional selection is practiced, the surviving families are expected to become more similar, both genetically and phenotypically, as less desirable families are eliminated. Effective selection may ultimately have the favorable effect of concentrating large numbers of favorable alleles within some families. Genetic assortative mating also increases genetic correlations among relatives within lines of descent (families). High genetic correlations among relatives are important in plant breeding because they are good measures of prepotency, which can be defined as the ability of individuals to impress their characteristics on their offspring. Full homozygotes are most prepotent because they produce only one kind of gamete, whereas individuals that are heterozygous at several loci are unlikely to be highly prepotent because they produce many kinds of gametes. As we have seen (Chapter 5), the most useful measure of homozygosity is the inbreeding coefficient, F. This coefficient takes a value near zero in most large random-mating populations, and the coefficient increases toward unity under sustained genetic assortative mating. Self-fertilization (one individual in each generation in each family) leads to very rapid increases in homozygosity. Starting with a heterozygote (F = 0.50), F takes the values 0.75, 0.875, 0.9375, 0.9688, 0.9844, 0.9922, . . . in successive generations of selfing, thus exceeding 0.90 in the third generation. Under continued mating of 2 individuals per family (full sibs) each generation, F is not expected to exceed 0.90 until the eighth generation. With continued mating of 4 individuals per family (double first cousins), F is not expected to exceed 0.90 until the seventeenth generation and rates of increase in F with 8 individuals per family (quadruple second cousins) and 16 individuals per family (octuple third cousins) are much slower yet. The rate of increase in F is so slow with more than 16 mating individuals per family per generation that such matings are of essentially no consequence in concentrating favorable alleles in selection programs in outcrossing species. It is therefore not at all surprising that breeders of plant species nearly always choose schemes featuring very close inbreeding. Selfing schemes (one parent/generation) are by far the most common in breeding outcrossing plants, and the usual goal is to develop numerous highly homozygous lines that are first evaluated by top crossing to identify lines with good "general combining ability," followed by testing specific combinations of pairs of lines to identify the very few pairs that have potential to produce truly excellent single-cross hybrids. The processes by which inbred lines are developed in outcrossing species, superior hybrids are identified, and truly outstanding inbreds are subsequently improved to produce still better hybrids, are discussed in Chapter 14.

## NONRANDOM MATING: PHENOTYPIC ASSORTATIVE MATING

It is intuitively obvious that the genetic effects of mating like-to-like in outcrossing species, on the basis of phenotypic appearance, depend on two factors: numbers of loci and heritability. If only a single diallelic locus is involved and no dominance or environmentally induced variations confuse phenotypic manifestations, $a_1a_1$ and $a_2a_2$ genotypes can be identified unambiguously. Mating of extreme types $a_1a_1 \times a_1a_1$ or $a_2a_2 \times a_2a_2$ can then be made with the following results: (1) achievement of homozygosity for $a_1a_1$ or $a_2a_2$ in a single generation; (2) concentration of individuals with extreme phenotypic expressions at one or the other side, or on both sides, of the phenotypic distributional scale in a single generation; (3) achievement of complete genetic correlations among members of a family in a single generation. If random mating is resumed among extreme types, the population is expected to return to its original composition, provided the selection practiced has not altered allelic frequencies.

The effect of mating like-to-like, without errors of classification, for a trait governed by two pairs of equally effective and equally viable alleles lacking dominance, in a random mating species, was worked out by Wright (1921). If the assortative mating involves only the two extreme phenotypes, fixation is achieved in one generation. But if it involves all possible phenotypes, there is a gradual decrease in the frequency of intermediate phenotypes and an increase in the two extreme phenotypes. Ultimately, all but the two extreme phenotypes will disappear. If heritability is not complete so that errors of classification occur, the ultimate result will be an equilibrium in which some intermediate phenotypes remain in the population. Just as in the case of one locus, mating of individuals that are alike in having extreme phenotypes results in concentration of individuals at the extremes, an increase in homozygosity, and an increase in the likeness of close relatives. The main difference between the one- and two-gene cases is the rate of progress toward the final outcome; progress is slower with two loci. The effect of further increases in the number of loci governing a character (heritability of the loci remaining constant) is increasingly slower progress. If heritability is complete for two loci, so that errorless assortative mating is possible, there can be an approach to fixation (F = 1); however, the approach to fixation will always be very slow unless the number of loci is small. When heritability is incomplete, leading to mistakes in classification, F cannot reach 1.

For many economically important characters, numbers of loci are likely to be large and heritabilities low. Consequently, mating like-to-like is not apt to have much effect until after a great many generations of selection, and even then it is unlikely to produce much fixation of type. It does, however, tend to increase the resemblance between close relatives, and it can be a useful tool in producing diversity within populations. Whereas genetic assortative mating tends to fix intermediate as well as extreme families, phenotypic assortative mating tends to shift the population toward the tails of the distribution curve and reduce the frequency of intermediate phenotypes. However, should random mating replace phenotypic assortative mating, these effects will rapidly be dissipated unless the selection accompanying the assortative mating has permanently altered allelic frequencies.

## NONRANDOM MATING: GENETIC DISASSORTATIVE MATING

Genetic disassortative mating finds little use in closed populations of the type we have been considering in this chapter. Its main application in plant breeding is a relatively recent development, namely, the crossing of unrelated strains with compensating strengths and weaknesses to develop populations with a broader genetic base that may be useful in future breeding projects.

## NONRANDOM MATING: PHENOTYPIC DISASSORTATIVE MATING

The greatest use of phenotypic disassortative mating is in maintaining genetic diversity in populations that are maintained as sources of alleles that affect desirable traits. By choosing and mating phenotypically unlike individuals in such populations, the breeder enhances the likelihood of maintaining genetic diversity. Furthermore, in contrast with assortative mating, disassortative mating tends to reduce genetic correlations among relatives (i.e., to increase heterozygosity) and to inhibit the separation of populations into genetically distinct subpopulations. Phenotypic disassortative mating is therefore a favored mating system, not only for developing source populations carrying alleles useful for breeding purposes, but also for long-term maintenance of such genetically enhanced populations. Phenotypic assortative mating is commonly regarded as the mating system that best maintains genetic variability within populations, including stocks that have been deliberately enhanced for characters of interest to breeders (e.g., high seed protein content, superior resistance to lodging, resistance to various diseases or pests). Such germplasm lines are registered monthly in the journal *Crop Science* for the convenience of breeders.

## GENERAL CONSIDERATIONS

Production traits, which are likely to be affected by alleles of many genetic loci, can also be much affected by environment. The effects of the five main sexual mating systems on such traits are summarized in the following paragraphs.

Random mating implies virtually unrestricted mating among all members of a population, leading to formation of hybrids among plants with differing genotypes. This in turn leads to extensive segregation and recombination that produces large numbers of new genotypes each generation. With effective selection, individual alleles as well as multilocus combinations of alleles with favorable effects on survival are expected to increase in frequency in the population, with the consequence that mean population fitness should also increase. However, the unrestricted outcrossing and the extensive segregation and recombination that follow tend to break up previously existing favorable combinations of alleles at a rate more or less equivalent to the rate of assembly of new favorable combinations by recombination. Thus, random-mating populations may reach equilibrium for single loci in one generation, but thereafter multilocus population structure develops slowly at best. With effective selection, however, population structure may adjust over time, slowly reaching closer agreement with an optimum population phenotype. The population can be expected to vary about the optimum structure from generation to generation with shifts in environment.

Close inbreeding without conscious selection rapidly causes previously random-mating populations to subdivide into numerous noninterbreeding, genetically distinct families. This almost always leads to so-called inbreeding depression in outcrossing species—a decrease in vigor and fitness which, although usually attributed to increased homozygosity of deleterious alleles, is more likely due to haphazard chaotic destruction of a favorable population structure that had previously built up under favorable environmental conditions.

Close inbreeding combined with effective natural selection automatically leads to linebreeding that is likely to have favorable effects in concentrating favorable alleles within selected families. However, such selection in outcrossing species will almost certainly be overmatched by the powerful destructive effects of segregation and recombination, and the outcome will nearly always be decreases in vigor and fitness as outbreeding populations separate in early generations into numerous increasingly homozygous and genetically distinct families. Such changes virtually cease after about generation 5 or 6 of selfing, when destruction of previously existing population structure approaches completion. At this point, crossing different plants within families nearly always fails to restore general vigor or fitness. However, different families (lines) are likely to be homozygous for different alleles of many different loci (especially lines separated early in the genetic assortative mating process or, more particularly, lines derived from different original populations) and crossing such lines nearly always leads to increases in vigor and fitness. In practice this has proved to be the case, and the breeding of hybrid varieties of outcrossing species now largely involves developing very large numbers of inbred lines, topcrossing to a common pollen parent (often an inbred-variety cross) to identify lines with superior general combining ability, followed by pairwise crossing of such lines to identify those rare pairs that have extraordinary specific combining ability. The operations involved in the breeding of superior hybrid varieties and their subsequent further improvement are discussed in more detail in Chapter 14.

## RANDOM MATING IN INBREEDING POPULATIONS

Random mating can also play a useful role in the breeding of self-pollinated species. This can be illustrated by the role played by random mating in developing a population that is proving useful in breeding for resistance to scald disease, caused by the fungus *Rhynchosporium secalis*, which is perhaps the most damaging disease of barley in Mediterranean-like climates. At least 30 different races (pathotypes) of this disease have been identified, but no single resistance allele has been discovered that provides adequate protection against all of 30 or more known pathogenic races (pathotypes of the fungus). This lack of any type with resistance to all pathotypes clearly called for a practical and simple methodical approach that would avoid assembling the numerous resistance alleles, one by one or in groups, that might provide protection against all of the various pathotypes that might be present in any given local area. Tests that had been made of accessions in the U. S. Department of Agriculture (USDA) world barley collection over many years had established that although no single barley accession provided protection against all races of scald, some were resistant to several races, and that within the entire array of resistance alleles combinations possibly exist that would protect against all existing races. Thus, provided that the proper alleles could be accumulated within a single barley stock, that stock would almost certainly be useful in breeding for general resistance. Accordingly, the following three-step process was devised and applied with the goal of avoiding the daunting task of attempting to incorporate resistance on a single race-by-race basis. The first step was to identify all Californian varieties of barley as well as all USDA accessions of barley that were known to be resistant to one or more races of scald. Thirty-three such varieties and accessions of barley were identified; it was hoped that these 33 items might, among them, carry alleles that would protect against all races of scald, or at least all important alleles of truly damaging races. The second step was to collect inoculum of the scald disease organism from all of the principal barley-growing regions of California; perhaps these collections would, among them, carry all important pathogenicity alleles needed to protect against all pathotypes of the scald organism present in California.

The third step was to isolate the 33 barley accessions with homozygous recessive male-sterile versions of three well-adapted and widely grown Californian commercial varieties of barley under conditions that would promote maximum intercrossing. The many seeds produced on the male-sterile plants clearly had to be crosses with one or more of the 33 scald-resistant accessions with which the male-sterile commercial varieties had been isolated. These seeds, which all led, as expected, to male-fertile $F_1$ hybrid plants, were planted to produce in turn a highly heterogeneous $F_2$ population made up of about 3/4 male-fertile and 1/4 male-sterile individuals. Seed set was apparently complete on the male-fertile plants but was also good on the male-sterile $F_2$ plants, thus providing for a second generation of extensive segregation in many individuals of the next generation. The following generation, which included $F_3$ individuals as well as newly formed first generation hybrids, was inoculated in the seedling stage, and at several later stages of development, with the scald inoculum from several locations in California, and all individuals with scald symptoms were rogued. The seeds from scald-free male-sterile plants were harvested in mass and seeded to produce a next generation of several thousand individuals. This process was repeated for two additional generations, at which time fertile and highly scald-resistant plants were identified. These plants were inoculated with scald at several stages during the life cycle, and again the increasingly smaller numbers of plants with scald symptoms were rogued. In the next generation virtually all plants were not only fertile but also failed to develop symptoms of scald on inoculation and, furthermore, produced seed yields that were closely equivalent to Californian commercial varieties. It was concluded that the procedures followed had been successful in transferring useful resistance to all, or virtually all, Californian races of scald into a genetic background that allowed for commercially acceptable levels of production. This scald-resistant population is presently being used as one parent in breeding scald-resistant barley varieties. Although the population procedure followed took nearly 10 years, little labor was required in any generation and the entire program was not labor intensive. Note, in particular, that the several generations of selfing provided opportunities for a number of generations of line breeding that efficiently improved general adaptedness while also fixing homozygosity for scald resistance.

# THIRTEEN

# *Breeding Self-Pollinated Plants*

During the late decades of the nineteenth century and early decades of the twentieth century a major shift occurred in the breeding of self-pollinated species. Early in this period observant breeders noticed that the occasional natural hybrids that occur between different pure lines often segregate to produce great variability in the $F_2$ and following early segregating generations. Encouraged that some pure-line segregants from such natural crosses appeared to be superior to the parental populations in which the hybrids had arisen, plant breeders began making artificial hybrids between different, apparently superior, pure lines and thereafter increasingly turned to selecting the "best" among pure-line segregants as sources of new varieties.

In the late nineteenth and early twentieth centuries nearly all plant breeders shifted to what is now often called the "classical pedigree method" in managing the segregating generations. In this method plants of the $F_2$ generation, grown from selfed seed of human-made $F_1$ hybrids, were space-planted, and individual superior $F_2$ plants were selected and allowed to self-pollinate to produce $F_3$ families. During generation $F_3$ to generation $F_5$ or $F_6$, superior single plants in families judged to be superior were selected and, as heterozygosity declined and between-family differentiation increased, family selection was increasingly substituted for single-plant selection within families. As families increasingly approached homozygosity, some closely related families were eliminated on the basis that they were likely to be similar in worth; testing for yield and quality were also likely to be started in those generations. In generations $F_7$ to $F_{10}$ or so, when only a few of the surviving families were likely to be descendants of any single original hybrid, intensive yield and quality testing were likely to be practiced; ultimately, the apparently single best-surviving family derived from any single $F_1$ was multiplied, named, and released for use in agriculture. Soon more and more breeders began to recognize that the classical pedigree method was excessively labor intensive and required burdensome record keeping. Further, the method featured substantial laborious evaluation of families in early generations when heterozygosity was high and heritability was low. Classical pedigree breeders also began to recognize that highly desirable families with the ultimate character they desired, "high overall worth," were often simply not present in the frequently small numbers of family lines that survived into generations $F_5$ to $F_8$. An obvious solution was to grow larger $F_2$, $F_3$, and $F_4$ populations, select on a single-plant basis in those generations, and delay family evaluation until homozygosity, and hence heritability, had greatly increased as a result of selfing.

## THE BULK-POPULATION METHOD

In present-day breeding of self-pollinated plants, the early segregating generations of crosses are usually managed by one or another of various bulk-population methods. Nilsson-Ehle of Sweden was perhaps the first breeder to use a bulk population method. He adopted this method because it appeared to offer advantages in handling the segregating generations of a hybrid he had made to combine the winter hardiness of the Squarehead variety with the high yield of the Standup variety of winter wheat. Nilsson-Ehle assisted natural selection in the $F_2$ generation by discarding plants that had (1) suffered winter damage but nevertheless had not been completely prevented from reproducing or (2) appeared to be susceptible to lodging or had other defects. He harvested single desirable plants, sowed their seeds in bulk, and continued single-plant selection into generations $F_5$ or $F_6$. He reasoned that growing large populations and eliminating clearly inferior plants in early generations would likely increase chances that a higher proportion of acceptable plants would appear among the winter-hardy and otherwise desirable selections. He believed that the bulk-population method was well suited to handling the large numbers he needed, because much of the selection in this method was on a single-plant basis, thus relieving him of the laborious and costly task of growing numerous progeny rows and keeping pedigree records in the early segregating generations. He further recognized that homozygosity would increase during the period of bulk handling, so that single-plant selections made after a few generations of bulk handling could be expected to breed nearly true.

The various bulk methods that have subsequently been proposed differ from one another in detail, particularly in respect to the numbers of single plants grown and the stringency of selection practiced in the early segregating generations. However, all present-day bulk population methods appear to share a common feature, namely, that selection among family lines is delayed until homozygosity is high and, hence, heritability is also high. Some breeders eliminate early generation selection entirely. Thus, in the single-seed-descent method progenies of $F_2$ plants are advanced as rapidly as possible by keeping only a single seed per plant through early segregating generations. Because only one seed is needed per plant to advance each generation, optimum plant development was deemed largely irrelevant, thus allowing these early generations to be grown under less than optimal environmental conditions—for example, in off-season nurseries or in greenhouse benches or cold frames in which appropriate temperature and lighting regimes could be used to force early flowering and early maturity. Thus, two, or even three, generations could sometimes be grown in a single year, thereby much reducing the time required to obtain the near-homozygous lines needed for more precise evaluation of "overall worth." Moreover, the overall worth of a new cultivar nearly always depends on the combinations of alleles of many loci, and there was general agreement that this characteristic can be determined reliably only by careful evaluation in more than one growing season of many plants of the ultimately emergent true-breeding or near-true-breeding lines. Which scheme is likely to be most efficient in providing the breeder, in the earliest possible generation, with arrays of adequately true-breeding families that are likely to allow identification of segregants with high overall worth? The sample of lines generated by the single-seed-descent method is expected to be a more or less random sample of the pure lines that any given $F_2$ hybrid is capable of producing. On the other hand, the sample of nearly pure lines that emerges from the various bulk-population methods will usually be a selected sample, and hence a superior one, provided that early generation single-plant selection has been effective. However, a pedigree method sample will have undergone more between-family selection in early generations and, if such selection is effective when heritability is low, this sample might be expected to include a higher proportion of superior lines than the sample obtained by the single-seed-descent method. A likely disadvantage of the pedigree method is that the number of surviving lines will be much smaller because fewer family lines can be accommodated in early generations than with the bulk or single-seed-descent procedures. We now examine the preceding assumptions, first in light of genetic theory, then in terms of evidence from experiments designed to test relevant assumptions, and finally in terms of the practices that have been adopted by present-day breeders.

## Genetic Theory

It is convenient to start with the simplest possible genetic situation, an $F_1$ hybrid heterozygous at a single locus for two adaptively neutral alleles ($a_1$ and $a_2$). Under complete selfing the expected genotypic composition of an $F_2$ population large enough to avoid significant drift is $1/4\, a_1 a_1 + 1/2\, a_1 a_2 + 1/4\, a_2 a_2$. Thereafter, one-half of the remaining heterozygosity is expected to be lost in each generation of complete selfing, so heterozygotes are expected to decrease according to the series $(1/2)^g$, in which $g = 0$ in $F_1$, $g = 1$ in $F_2$, and $g = 7$ in $F_8$. The proportion of homozygotes is thus expected to increase according to the series $1 - (1/2)^g$, whereas the probability of heterozygosity ($h$) after $g$ generations will be $h = (1/2)^g$ and the probability of homozygosity will be $P(h) = [1 - (1/2)^g$. Hence, after $g$ generations the probability of heterozygosity $P(h)$ at each locus will be $P(h) = 1.00, 0.50, 0.25, 0.125, 0.0625, 0.0313, 0.0156,$ and $0.0039$ in generations $F_1$ ($g = 0$), $F_2$ ($g = 1$), . . . $F_8$ ($g = 7$). Correspondingly, the series for homozygosity will be $P(H) = 0.00, 0.50, 0.75, 0.875, 0.9375, 0.9687, 0.9844,$ and $0.9961$ in generations $F_1$ through $F_8$. These inverse trends for decreasing heterozygosity and increasing homozygosity are often represented in terms of genotypic frequencies ($a_1 a_1 : a_1 a_2 : a_2 a_2$); for example, in $F_2$ expected frequencies are $0.25:0.50:0.25$ and in $F_8$ they are $0.4961:0.0078:0.4961$. Thus, in the single-locus case with two alleles, the heterozygote ($a_1 a_2$) is expected to decrease rapidly in frequency and the two possible homozygotes to increase rapidly in frequency. The situation is similar but more difficult to visualize when 2 or, especially, more loci are considered simultaneously over a number of generations. If all loci behave completely independently (no interactions of any kind among loci) the probability of heterozygosity is $(1/2)^g$ and the probability of homozygosity for

1 locus after $g$ generations of selfing is given by $P = [1 - (1/2)^g]$. The computations are tedious, but preparing a table, such as Table 13-1, that gives expected homozygosity values >0.01 and <0.99 for selected generations $g = 0$ to $g = 13$ ($F_{14}$) and for $n = 1, 2, 3, 5, 10, 20, 50, 100,$ and 150 loci permits preliminary assessments of the possible worth of the segregants that appear over generations.

Table 13-1 shows that if an $F_1$ is heterozygous for 10 loci that have a bearing on overall worth, slightly more than half (0.77) of the selfed lines descended from such an $F_1$ plant are expected to be entirely homozygous by generation $F_5$ (many more will be nearly homozygous) and about 92% of selfed lines are expected to be entirely homozygous by generation $F_8$ (most of the remaining 8% are also expected to have approached homozygosity quite closely). This suggests that adequate levels of homozygosity may have been reached by generations $F_5$ to $F_8$ and, hence, that initiating between-line selection in generation $F_5$ might be rewarding. Contrariwise, if an $F_1$ is heterozygous for 50 loci that have a significant bearing on overall worth, only 4% of $F_5$ lines are expected to be fully homozygous by generation $F_5$ and only 68% by generation $F_8$. This suggests that the effectiveness of between-line selection may be quite limited, at least until generation $F_8$, and that single-plant selection may provide additional improvements in estimating the probable value of single plants at least into generations $F_5$ to $F_8$. However, in the absence of reliable information about the numbers of loci involved (or estimates of the magnitudes of favorable or unfavorable effects of different alleles, or estimates of the magnitudes of the effects of different loci, or information concerning the effects of interactions among loci), speculations concerning the duration of gain under selection are unlikely to provide reliable guidance for breeders. Another and possibly more important consideration is the likelihood that some alleles with negative effects may be lost early, perhaps as early as generation $F_2$. For one of a pair of neutral alleles, the likelihood of fixation of one or the other allele of the pair is $1 - (3/4) = 0.25$, or 25%, in $F_2$; however, if one allele of each pair is superior to the other, the likelihood of loss is expected to be greater for the poorer and the likelihood of fixation greater

TABLE 13-1   Values of $P = [1 - (1/2)^g]^l$ for representative generations ($g$) and representative numbers of loci (l) (see text)

| | GENERATIONS ($g = 0$ IN $F_1$ AND $g = 13$ IN $F_{14}$) | | | | | | | | | | | | |
|---|---|---|---|---|---|---|---|---|---|---|---|---|---|
| LOCI | $F_2$ | $F_3$ | $F_4$ | $F_5$ | $F_6$ | $F_7$ | $F_8$ | $F_9$ | $F_{10}$ | $F_{11}$ | $F_{12}$ | $F_{13}$ | $F_{14}$ |
| 1 | 0.50 | 0.25 | 0.88 | 0.94 | 0.97 | 0.98 | 0.99 | — | — | — | — | — | — |
| 2 | 0.25 | 0.56 | 0.77 | 0.88 | 0.94 | 0.97 | 0.98 | 0.99 | — | — | — | — | — |
| 3 | 0.125 | 0.42 | 0.70 | 0.77 | 0.91 | 0.95 | 0.98 | 0.99 | — | — | — | — | — |
| 5 | 0.0313 | 0.24 | 0.51 | 0.72 | 0.85 | 0.92 | 0.96 | 0.98 | 0.99 | — | — | — | — |
| 10 | — | 0.06 | 0.26 | 0.52 | 0.77 | 0.85 | 0.92 | 0.96 | 0.98 | 0.99 | — | — | — |
| 20 | — | — | 0.07 | 0.28 | 0.53 | 0.73 | 0.86 | 0.92 | 0.96 | 0.96 | 0.99 | — | — |
| 50 | — | — | — | 0.04 | 0.20 | 0.46 | 0.68 | 0.82 | 0.90 | 0.95 | 0.98 | 0.99 | — |
| 100 | — | — | — | — | 0.04 | 0.20 | 0.46 | 0.68 | 0.82 | 0.90 | 0.95 | 0.98 | 0.99 |
| 150 | — | — | — | — | 0.01 | 0.05 | 0.31 | 0.56 | 0.74 | 0.86 | 0.93 | 0.97 | 0.99 |

— Indicates frequency < 0.01 or > 0.99.

for the superior allele of any locus. For 2, 5, 10, and 50 pairs of neutral alleles the likelihood of fixation, as early as $F_2$, of at least some alleles increases rapidly, and with larger numbers of generations the likelihood increases effectively to 1.00 after 10 or more generations. Thus, unless the number of loci is very small, it seems unlikely that any individuals with the potential to produce the very best possible segregant will survive into the sample of near-homozygous lines present in generations $F_5$ to $F_8$, the generations in which intensive evaluations for overall worth usually begin. One obvious countermeasure is to grow the largest feasible $F_2$ population and to eliminate in generations $F_2$ through $F_5$ to $F_6$ all individuals with visible defects or unimpressive phenotypes. Bulking the progeny of the remaining plants and repeating single-plant selection requires relatively small effort; hence, the attractiveness of such single-plant selection is obvious. If the survivors include only a somewhat higher proportion of plants with improved potential than are present in generation $F_2$, the prospects of coming closer to the ultimate limit of gain should be enhanced. However, few breeders are so optimistic as to believe such selection could possibly lead to anything approximating full fixation of all of the favorable alleles that trace back to any single $F_1$ hybrid combination, or to elimination of most of the poorer alleles.

Another feasible and possibly effective countermeasure is to make second-cycle hybrids between the very "best" appearing individuals that survive into generations $F_5$ to $F_8$. It seems unlikely that many lines derived from first-cycle hybrids would carry precisely the same favorable alleles or that after several generations of natural selection, perhaps aided by breeder selection, they would be deficient in precisely the same unfavorable alleles. Hence, second-cycle or later-cycle hybrids might well provide opportunities to achieve higher levels of performance either through accumulation of additional favorable alleles or, more likely, by repairing defects caused by deleterious alleles. Although there have been anecdotal reports that second-cycle and later-cycle hybrids do indeed provide opportunities to improve overall worth, there is no fully convincing information in the literature that this is indeed the case. It therefore seems worthwhile to report the results of some experiments that were designed to determine the effectiveness of differing methods of managing early generation breeding materials.

## Experimental Evidence: Survival in Mixtures

In perhaps the earliest experiment to test the effectiveness of natural selection in eliminating inferior genotypes, Harlan and Martini (1938) made a mixture containing equal numbers of seeds of 11 phenotypically distinctive barley varieties and grew the mixture for various numbers of years at 10 environmentally distinctive locations in the western and northern United States. Each year the mixtures were harvested in mass, and random samples of seeds were used to establish two plots the following year, one for the purpose of taking a census of the frequency of each variety remaining in the mixture and the other to seed the next generation. The 11 varieties in the experiment included some varieties that were widely adapted, some varieties that were poorly adapted nearly everywhere, and some with specific and narrow adaptedness. The most striking result of the experiment was the rapidity with which only a single variety became dominant at nearly all locations, accompanied by elimination or near elimination of all or most other varieties. Changes occurred slowly at only a very few locations, and two or more varieties persisted in moderate frequencies at three locations.

These results showed that natural selection was unquestionably a significant force at all locations; they also answered several other questions. First, did the changes in frequency follow theoretically expected changes? Theoretically, the most competitive genotype should ultimately dominate any population, and there was, in fact, no question as to which variety became dominant at all, or nearly all, of the 10 locations. Harlan and Martini also found many instances of curves of rapid elimination; at least one variety was eliminated rapidly at all 10 locations, and in each case the variety eliminated was clearly recognizable as quite inferior at that location. At some locations nearly all of the elimination curves were of this type. Furthermore, the curves for mediocre varieties are expected to fall into two classes. The poorer of such mediocre types is expected to decrease at a rate represented by steadily decreasing straight lines, and the better types by humped curves (early increase in frequency when most of the competition is against poorer types, but later by a declining more or less straight line when most of the competition is expected to be against the better competitors); many examples of both types of curve were found. This experiment also provided useful information about the relationship between competitive ability and agricultural value. Coast and Trebi were clearly the outstanding commercial varieties in the mixture. These two varieties made a poor showing at only one exceptionally arid location, where they were eliminated by White Smyrna, a variety noted both for exceptional adaptedness and for high yielding ability under such highly arid conditions. Hannchen, a widely adapted commercial variety with good yielding ability, made a reasonably good showing at all locations. Furthermore, the dominant variety differed from the variety usually grown by farmers in the same vicinity at only two locations; also in these two locations, the preference by farmers for a variety other than the dominant variety of the region was based on considerations other than yield. The agreement between superiority, mediocrity, and/or distinct inferiority and survival was even more striking. All of the mediocre varieties declined in frequency, clearly inferior varieties were rapidly eliminated at all locations, and all clearly superior varieties became predominant.

### Competitive and Cooperative Interactions

The preceding informative experiment stimulated several other similar experiments from which it became apparent that a simplification made in earlier models of populations that were changing in time and space—namely, that the selective values of the competing genotypes are constant—is often not valid. When different individuals coexist in a population, it is possible to imagine a number of competitive and cooperative interactions that can probably be illustrated most simply in terms of experimental data. Table 13-2 gives the results of an experiment in which four phenotypically distinctive commercial barley varieties (Atlas, Club Mariout, Hero, Vaughn) were grown in pure stand and in competition with each other. The values

TABLE 13-2   Yield of four commercial barley varieties in competition with specific other varieties expressed as a percentage of the yield of each variety in competition with itself

| VARIETY | ATLAS | CLUB MARIOUT | HERO | VAUGHN | MEAN |
|---|---|---|---|---|---|
| Atlas | — | 103.1 | 102.6 | 101.4 | 102.4 |
| Club Mariout | 106.0 | — | 104.0 | 103.0 | 104.3 |
| Hero | 101.8 | 101.8 | — | 102.1 | 101.9 |
| Vaughn | 96.3 | 100.4 | 102.0 | — | 99.6 |
| Mean | 101.4 | 101.8 | 102.9 | 102.2 | 102.0 |

The rows in this table give the yield of each variety when it was surrounded by specific other varieties and, hence, indicate competitive ability. The columns indicate the effect of each variety as a competitor on the yields of its neighbors. Vaughn was the only poor competitor, and all of the varieties were, on average, good neighbors (from Allard and Adams 1969).

in this table represent the yield of each variety in competition with a specific other variety, expressed as a percentage of yield of each variety in competition with itself. It is important to note that yields in competition were larger than pure-stand yields in 11 of the 12 cases. The relationships among these varieties are thus more often appropriately described as those of cooperation or enhancement rather than of competition. The rows in Table 13-2 give the yield of each variety when it was surrounded by specific other varieties and, hence, indicate competitive ability. The columns indicate the effect of each variety as a competitor on the yields of its neighbors. Mean values show that Vaughn was the only poor competitor (mean performance < pure-stand performance) among these four varieties and that, on average, all of the varieties were good neighbors (mean performance > pure-stand performance). Although the yields of these four varieties were rather consistently higher in mixture than in pure stand, the yield increments in mixtures were always rather small (usually < 5%). This is consistent with the results of many other experiments involving mixtures of commercial varieties; it is hardly expected that varieties selected for high yielding ability in pure stand would have biological properties conducive to favorable interaction in mixture. It seems much more likely that genotypes with such properties would be found in populations with a history of mutual selection. This brings us to a complex area of research that encompasses elements of population and ecological genetics as they impinge on evolutionary outcomes of mixtures of homozygotes in populations, and on breeding practices.

## HETEROZYGOSITY, HETEROGENEITY, AND POPULATION STRUCTURE

The simplest proposal concerning the overall performance of a heterogeneous population is that its performance can be predicted from the sum of the independent performances of its components. This assumption is frequently made, but as shown in the Harlan and Martini experiment and in the experiment with four barley varieties, Atlas, Club Mariout, Hero, and Vaughn, discussed in the preceding sections, it is often not justified. A large number of studies, mostly of simple physical mixtures of pure lines of inbreeders (e.g., barley, wheat, rice, *Phaseolus* beans) have been carried out. In these experiments mixtures have nearly always yielded more than the mean of their components and are often more stable (homeostatic) in performance than their components grown in pure stand. This suggests that heterogeneity often leads either to (1) phenotypic interactions that provide gains in performance and/or (2) mutual buffering or homeostasis that results in steadier performance. Moreover, substantial gains in performance and homeostasis have often been observed in studies of complex experimental populations of highly successful natural populations of successful, almost completely self-pollinating species, as well as in complex populations of heavily self-pollinating crop plants (e.g., *Phaseolus* beans, barley, and rice) (Allard 1965). Such populations simply do not go to fixation over many generations. Even though these populations were heavily self-pollinated, large numbers of distinctly different genotypes often persisted in them for many generations—sometimes indefinitely. The usual explanation advanced to account for the persistent heterozygosity is true overdominance. But this explanation is simply not consistent with the low levels of outcrossing (1 or 2%) actually measured in such populations using marker loci. The substantial genetic diversity observed in all or nearly all such populations is also often not consistent with strict intergenotypic competition. Under straightforward competition, the single best competitor soon wins (or reaches high frequency) and all or nearly all of the poorer competitors are soon eliminated or reduced to low frequency. It seems much more likely that selection favoring genotypes that are favorably cooperative (synergistic) neighbors may also provide an explanation for the population structures that actually develop.

To test the aforementioned hypothesis, an experiment was carried out with eight genotypes chosen at random from the 18th generation of barley Composite Cross V (CCV). CCV was synthesized from a mixture of equal numbers of $F_2$ seeds from each of the $(N)(N-1)/2 = 465$ possible $F_1$ intercrosses among 31 barley varieties of diverse geographical origins; thereafter CCV was maintained without conscious selection. In the early generations the yield of CCV was about two-thirds that of the most popular Californian commercial varieties. However, by generation 15 the yield of CCV had increased to 90% or more of the most popular commercial varieties of the time and thereafter the yield of the population continued to increase, but even more rapidly. The population remained highly diverse genotypically into generation $F_{45}$, when it was last examined in detail regarding the property of diversity. By far the most interesting feature of this experiment was the contrast between the results with mixtures of commercial varieties and results obtained with genotypes derived from CCV when such genotypes were tested in an experiment parallel in design to that reported in Table 13-2. Although the commercial varieties generally produced larger yields in mixture than in pure stand, the yield increments associated with such varieties in mixtures were always quite small and statistically insignificant. In sharp contrast, yield increments were larger and often statistically significant for most of the genotypes that had been isolated from CCV. The commercial varieties had been developed in breeding programs designed to identify types with superior performance in pure stands, and it is hardly to be expected that varieties thus selected would have biological properties conducive to synergism. The genotypes from CCV, on the other hand, had a history of large numbers of generations of selection in competition within large and diverse arrays of genotypes. It is not surprising that selection in such mixed populations favored the survival of genotypes that have superior "ecological combining ability," in the sense that they were good neighbors and at the same time good competitors. It is particularly interesting that 2 (25%) of the 8 genotypes isolated from CCV for detailed testing were at least equal to, and probably superior to, good-yielding commercial varieties and that only 3 isolates were inferior (but not significantly so). Additional evidence that favorably interacting genotypes developed and survived in CCV was provided by yields of a random sample of 29 genotypes isolated from a late generation ($F_{40}$) of a similar population (CCII). Pure-stand yields of 20 of the 29 genotypes were higher (6 significantly higher) than those of good-yielding commercial varieties. Moreover, the mean yield of a mixture containing equal proportions of the 29 components was ~11% higher than the mean yield of the 29 components grown separately. Natural selection for superior "ecological combining ability" is apparently commonplace in mixed populations; however, and perhaps more interesting to breeders, natural selection in competition also appears to perpetuate many survivors with superior pure-line performance.

## MODELS

The foregoing experiments show that only certain varieties or genotypes are superior when grown in specific associations with other genotypes; this was especially the case for genotypes isolated from complex evolving populations such as CCV and CCII. However, it was not obvious which genotype(s) would eventually attain preeminent places in such populations. It was therefore of both practical and theoretical interest to explore the consequences of evolution in such populations over various time intervals, because such information seems likely to be of value in guiding breeders. Unfortunately, direct experimental resolution of many of the issues involved is extraordinarily difficult owing to the large numbers of parameters, genetic as well as ecological, that are involved in genetic change in complex populations. Another potentially troublesome feature is the probable moderately long-term nature of the kinds of experiment required. Fortunately, in populations of this type the several factors that jointly affect survival of any particular homozygous genotype for two successive generations can be summarized in a single numerical value that relates frequency in any given generation $N$ to its frequency in the next generation (generation $N + 1$). These values, and their fluctuations from generation to generation, can be estimated quite precisely from census data for each distinguishable genotype and analyzed in both deterministic and stochastic models to predict the outcome of evolutionary processes in actual as well as hypothetical populations (Allard and Adams 1969).

Four types of interaction between any two genotypes, $X_i$ and $X_j$, are of particular interest: (1) neutral interactions ($K_{j/i} = K_{i/j} = 0$), in which $K_{j/i}$ is the difference in reproductive capacity of genotypes $X_i$ and $X_j$ in pure stand and in mixture, implying that neither genotype is affected by competition; (2) undercompensatory interactions $(K_{j/i}+K_{i/j}<0)$, implying competition to mutual disadvantage; (3) complementary interactions $(K_{j/i}+K_{i/j}=0)$, implying that gain in one combination is offset by loss in the other; and (4) overcompensatory interactions $(K_{ji}+K_{ij}>0)$, implying cooperation to mutual advantage. When there are more than two genotypes, numbers of possible higher-order interactions increase rapidly. However, experimental data show that higher-order interactions are often very complicated and nearly always too small to be of significance; hence, unless there is compelling evidence to the contrary, only first-order interactions between genotypes $X_i$ and $X_j$ need be considered.

Neutral interactions between pure lines of equal reproductive capacity have the same properties as Hardy-Weinberg gene models (no selection, no interaction), and they lead to similar expectations, namely, that genotypic frequencies will remain constant, but if disturbed for some reason, they again remain constant but at the new frequencies. Experimental data clearly establish that differences in reproductive capacity are nearly always a reality in actual populations and that such differences are involved in changes in genotypic frequencies. As discussed earlier, the frequency curves of the best and the poorest genotypes are monotonic, leading ultimately to fixation or elimination. However, the curves of the better intermediate types are humped, because frequencies of such types are expected to increase until types with higher reproductive capacity exceed them in frequency. If the population is small or if the reproductive values fluctuate randomly, there will be an array of outcomes among which the most likely will be the deterministic outcome.

The second type of system, involving undercompensating interactions, has in common with genetic systems featuring underdominance in fitness (heterozygote disadvantage), the property of lack of stable nontrivial equilibrium points. In such systems competition with other genotypes leads to reductions in reproductive capacity. If pure-stand reproductive capacities are all equal or very nearly equal, the genotype originally present in highest frequency ultimately dominates the population, even though it is in no way superior to other genotypes. Nevertheless, such systems become more conservative as the number of genotypes with nearly equal reproductive capacity increases, with the result that it may take a long time for fixation to occur when numerous individuals with more-or-less equal undercompensatory reproductive interactions compete against one another. In general, however, reproductive capacities will not be equal nor interactions balanced; if either of these contingencies turns out to be the case, both poor yielders as well as poor competitors are expected to be eliminated, and often the elimination will be rapid. This is, perhaps, the reason that actual populations have been found to contain so few genotypes that interact unfavorably. Thus, both experiment and theory lead to the conclusion that undercompensating interactions are even less helpful than neutral systems in explaining the omnipresence of the great genotypic diversity that has been found in nearly all real populations.

The third type of interaction, the type in which losses in some combinations are offset by gains in other combinations, is somewhat more helpful in explaining the diversity of genotypes found in real populations. This is because some patterns of complementary interactions, at least in theory, allow more than one genotype to remain permanently in populations. In such systems the poorest competitor is expected to be eliminated rapidly, but the better competitors may remain in populations permanently. Frequencies do not, however, converge on stable equilibrium values, but oscillate endlessly. Even though some patterns of complementary interaction do, in theory, allow many genotypes to persist for many generations, the infrequency of complementing genotypes that have been observed in real populations suggests that this type of interaction is unimportant.

The fourth interaction type, overcompensation, is by far the most interesting for two reasons: (1) because genotypes that show cooperative interactions have been found to be very common in populations that have a history of mutual selection and (2) because overcompensation can be shown to be a necessary condition for truly stable equilibrium. Overcompensating systems have properties similar to those of genetic systems, involving advantage of heterozygotes over homozygotes, and properties similar to those of favorable epistatic interactions between or among different loci; in particular, both types of interaction system often lock onto a stable equilibrium point.

## PREDICTED AND OBSERVED PERFORMANCE

Various competitive effects and/or enhancing effects, acting alone or in combination, obviously may lead to a wide variety of outcomes, and it is of interest to compare theoretically possible effects with observations of actual populations. An experiment involving four pure-line commercial barley varieties, Atlas, Club Mariout, Hero, and Vaughn, established that interactions had effects that were nearly always smaller than the differences among pure stand yields. This suggested that a population synthesized from these four varieties should follow the rules of neutral systems, that is, that the highest-yielding variety (Atlas) should soon dominate, forcing the three other varieties from the population. In an actual equal mixture of these four varieties Atlas (the best-yielding variety) reached its predicted frequencies of approximately 50% by generation 6 and reached a frequency of more than 90% by generation 13; thus, as predicted by theory this population was apparently well on its way to fixation of the variety Atlas by generation 13. Composite Cross V represents a much more complicated situation than the four-variety mixture just considered. This is because, as shown experimentally by competition experiments, CCV includes large numbers of genotypes that differ widely in both pure-stand reproductive capacity and in intergenotypic interactions. Although the eight genotypes isolated from CCV that were studied in a competition experiment parallel to that discussed previously represent only a microcosm of the complexity of the actual CCV population, it nevertheless seemed worthwhile to attempt to predict the outcome of competition among these eight genotypes, using simulation models such as those described earlier. Computer simulations showed that favorably interacting genotypes 7 and 3 were expected to become dominant rapidly and to come into stable equilibrium with each other within 25 to 50 generations. Genotype 7 had the highest pure-stand yielding ability of the eight genotypes studied (about equal to Atlas and 11% superior to the commercial variety Club Mariout) but was the poorest competitor. Genotype 3, in contrast, was a spectacularly good competitor but was the poorest yielder among the eight genotypes (about 4% inferior to Club Mariout). From the standpoint of plant breeding, it is particularly interesting that a population made up of these eight varieties (genotypes) was predicted by computer simulation of the selection model described earlier to come into an equilibrium made up of about 60% of genotype 7 and 40% of genotype 3 sometime between generations 25 to 50; the computer prediction that the yield of this two-genotype population would be about 4% greater than the pure-stand yield of the higher-yielding component (genotype 7) was also encouraging.

Estimates of the reproductive values on which these predictions were based have considerable errors of estimation. To estimate the effects of such errors, stochastic predictions were also made, in which the reproductive values were varied upward or downward by as much as 2 or 3 standard errors. Such changes had little effect on outcomes; this suggested that the evolution of any population was likely to be determined primarily by deterministic elements and that the evolution of such populations might be little affected by year-to-year environmental fluctuations. Note that outcrossing populations are much more complex genetically in all generations than inbreeding populations and hence are much less susceptible to precise analyses.

From the standpoint of plant breeding, perhaps the most appealing feature of mixed mass populations as a possible breeding procedure was the discovery that intergenotypic interactions can lead to stable feedback systems—systems that promote retention of large numbers of favorably interacting genotypes in heavily inbreeding populations—thus setting the stage for continued release of genetic variability as occasional natural crosses among the diverse genotypes within such populations infuse $F_1$ hybrids, as well as infuse continuing infusions of superior segregants from such hybrids into the population. Perhaps this is the reason that most of the originally mixed populations that have been studied have become much more complex. Thus, segregation and recombination, moderated by linebreeding in selfing populations, may be able to do much of breeders' work for them. In short, mixed mass populations in selfers appear to provide a method for exploring genetic variability inexpensively and on a scale not possible with outcrosses (Chapter 12). In this exploratory process intergenotypic interactions appear to be especially important, owing to the key role they appear to play in the survival, in mixed populations, of superior genotypes that promote long-term progress.

We now turn to a study that asked whether natural selection alone, or natural selection supplemented by breeder selection, is more efficient in exploiting the potential of additional cycles of natural intercrosses that occur among segregants from hybrids among successful varieties of barley, a heavy inbreeder.

## NATURAL SELECTION VERSUS
## BREEDER-SUPPLEMENTED SELECTION

Three barley varieties, Atlas, Club Mariout, and Vaughn, were chosen for this study because they are related in ancestry and because they are all popular with farmers and widely grown. The study was designed to answer two specific questions: (1) the extent to which later-cycle hybrids might improve the performance and/or repair defects among segregants from the already well-worked gene pool represented by the three varieties and (2) whether natural selection supplemented by breeder selection is more efficient than natural selection alone in advancing performance. There have been many reports in the literature that $F_1$ hybrids between different homozygous lines of selfers are higher yielding than their inbred parents, but that in $F_2$ and early segregating generations there is usually a decline in yield. Thus, in an often quoted study by Immer (1941), the average yields of six barley varieties, the six $F_1$ crosses among these 6 parent varieties, and four later selfed generations ($F_2, F_3, F_4$) of the six crosses were (expressed in percentages) as follows: parents 100, $F_1$s 127, $F_2$s 124, $F_3$s 113, and $F_4$s 105. However, a cautionary note has often been sounded regarding such studies, namely, that $F_1$ yield measurements may be suspect owing to the "necessity" for space-planting the scarce expensive hand-pollinated $F_1$ hybrid seeds of most selfers in yield trials, as well as the possible substantial advantage that the abnormally large $F_1$ hybrid seeds, borne on poorly set hand-pollinated spikes of barley, might confer on the $F_1$ plants grown from such seeds. In the experiment now to be considered, the three parental varieties were planted with seeds equivalent in size to those of the hand-pollinated $F_1$ seeds. These larger parental seeds were obtained by screening large seed lots of the parental varieties to separate larger from smaller seeds. This screening was successful in that the seeds used to sow the yield trials were closely equivalent in size to those of the hand-pollinated $F_1$ hybrid seeds. The larger parental seeds, as well as the hand-pollinated $F_1$ seeds, were sown at rates equivalent to farmers' rates of seeding in replicated yield trials in two successive years (five replications per year). The mean seed yields of the Atlas × Vaughn, Atlas × Club Mariout, and Club Mariout × Vaughn $F_1$ hybrids were not significantly different from each other, but the $F_1$ seed yields of the three hybrids were considerably lower than the $F_1$ hybrid yields that have usually been reported in the literature.

The $F_1$ hybrids between these three pairs of varieties all exceeded their respective parental mean yields by nearly 18%, whereas the usual yield advantage of $F_1$ hybrids over parental means reported in the literature has been about 25% to 30%; the 18% advantage observed in this study nevertheless indicates a substantial and statistically highly significant yield advantage of the $F_1$ plants over their no less phenotypically uniform parents. The three $F_2$ populations were much more variable phenotypically than the $F_1$ generation, and this was also the case in generations $F_3$ through $F_{13}$. The average yields of the three hybrid barley populations in the several later generations (in percentages of the parental means) were as follows: parental means 100, $F_1$ 118, $F_2$ 113, $F_3$ 105, $F_4$ 102, $F_5$ 101, $F_6$ 100, $F_7$ 103, $F_8$ 106, $F_9$ 108, $F_{10}$ 108, $F_{11}$ 109, $F_{12}$ 109, $F_{13}$ 110. The decreases observed during generations $F_2$ to $F_3$, $F_3$ to $F_4$, and $F_4$ to $F_5$ were not surprising. Evidently, they were attributable to the sharp decreases in propitious epistatic effects that are expected during the early segregating generations; destructive segregation almost certainly introduced many inferior genotypes into generations $F_2, F_3, F_4$, and $F_5$. Apparently, this decreasing trend in population performance largely ended in generations $F_5$ to $F_7$, when destructive segregations probably came to an end and the effects of the assembly and the introduction of advantageous epistatic multilocus genotypes owing to linebreeding likely began to have a favorable influence. Further support for this argument came from continuing increases in yield during generations $F_6$ through $F_{13}$. Electrophoretic assays made at two-generation intervals from generation $F_2$ through generation $F_{13}$ also support this argument; many of the 2-locus electrophoretically detectable allelic combinations that had been present in the parents and in the generation $F_1$ hybrid were again found in generations $F_8$ through $F_{12}$. Nevertheless, yields changed little from generation $F_9$ to $F_{13}$, and the frequency of clearly inferior plants, especially tall late plants subject to lodging, remained high during that five-generation interval. Population changes were much slower than had been the case in the earlier study, in which competition had been among only four well-adapted varieties, Atlas, Club Mariout, Hero, and Vaughn. This suggested that the problems that developed during generations $F_9$ to $F_{13}$ might have been that most of the competition in those generations was with plants that were clearly inferior agriculturally; if this was so, the removal of such plants from the population might more closely duplicate the sorts of competition that had prevailed in the earlier encouraging experiment with Atlas, Vaughn, Hero, and Club Mariout. Accordingly, the experiment was continued from generations $F_{14}$ through $F_{20}$, during which period rigorous selection was practiced against obviously inferior types in generations $F_{14}$, $F_{15}$, and $F_{16}$. Thereafter the survivors, mostly desirable agriculturally, were allowed to compete in generations 17, 18, 19, and 20, during which yields increased steadily, ultimately reaching a population mean slightly higher than the original $F_1$ yield (118%) of the three parents. It is not known how much of this yield increment was due to favorable interactions at the phenotypic level among the components of the population; clearly, not all was due to this factor, because among the pure lines that were isolated from the mixture, some were equal in yielding ability to the mixed population. The three experiments described in this chapter thus suggest that segregation and recombination, aided by natural selection, sometimes improve yielding ability either by (1) repairing defects in segregants from already well-worked gene pools or (2) increasing the frequencies of pure lines that interact favorably with each other at the phenotypic level to give higher yields in mixture than in pure stand.

## BACKCROSS BREEDING

The backcross method of breeding has been widely used to improve homozygous varieties and breeding stocks, as well as inbred lines of outcrossing species that are generally superior in overall worth but deficient in one or only a few specific characteristics. As the name implies, the method calls for a series of backcrosses to a generally superior variety or stock, during which some particular character to be improved is maintained by selection. Beginning in the $F_1$ generation and continuing for several generations, hybrid plants with the allele (or alleles) to be substituted are successively backcrossed to the parent to be improved. Thus, the generally well-adapted parent (the recurrent parent) into which an allele is to be substituted is involved in each backcross; the other parent (the donor parent) is involved in only the original cross. At the end of the backcrossing the allele (or alleles) being substituted will be heterozygous. Selfing after the last backcross produces homozygosity for the allele (or alleles) being substituted and, coupled with selection, will result in a variety (or stock) with exactly, or very nearly exactly, the adaptedness, yielding ability, and quality characteristics of the recurrent parent but superior to that parent in respect to the particular characteristic(s) for which the improvement program was undertaken. It is apparent that this method of breeding, in contrast with virtually all other methods, provides plant breeders with a unique degree of control of allelic frequencies in their populations.

Although line breeding, which features repeated crossing in a line derived from a common ancestor, has been used by animal breeders for centuries in attempts to fix breed characteristics in domestic animals, the great potential of backcross breeding in plant breeding seems to have escaped attention until Harlan and Pope (1922) pointed out the possibilities of the method in improving self-pollinated small grains. They observed that the backcross method had been "largely if not entirely neglected in definite programs to produce progeny of specific types." They suggested the probability that there were many instances in which backcrossing would be of greater value than the more common procedure of selecting during a program of selfing following hybridization. It was also in 1922 that Briggs (1938) started an extensive and long-continued program to add resistance to various diseases to well-established varieties of self-pollinated cereals. Briggs emphasized that "within certain limits the method was scientifically exact because the morphological and agricultural features of the improved variety could be described accurately in advance, and because the same variety could, if desired, be bred a second time by retracing the same steps."

## The Genetic Basis of Backcross Breeding

The $F_2$ generation obtained by selfing the $F_1$ of the cross $AA \times aa$ is expected to be made up of $1/4$ $AA$:$1/2$ $Aa$:$1/4$ $aa$ individuals. Even though half of the individuals are homozygotes, only $1/2$ of these homozygotes—or $1/4$ $F_2$ of the total population—will be of the superior and desired genotype of the recurrent parent, say $AA$. However, if instead of selfing the $F_1$, it is backcrossed to the recurrent parent, the expectation in the first backcross is $1/2$ $AA$ and $1/2$ $Aa$ individuals. Thus, in the backcross, half of the progeny are expected to be desired $AA$ homozygotes. The same expectation holds for each gene pair by which the two parents differ. If additional backcrosses are made to the recurrent parent, the hybrid population progressively becomes more like that parent instead of breaking up into $2^N$ homozygous genotypes as with selfing. If, for example, the parents differ by 10 gene pairs and no selection is practiced, six backcrosses will produce a population in which 99.4% of individuals are expected to be homozygous and identical with the recurrent parent at all 10 loci; that is, virtually all alleles from the donor parent will have been eliminated after six backcrosses. The rate at which alleles that enter a hybrid from the donor parent are eliminated during backcrossing will, of course, be affected by linkage. Suppose, in a backcrossing program undertaken to transfer the desirable allele $A$ from a generally undesirable parent (the donor parent) to a generally superior variety (the recurrent parent), that an undesirable allele $b$ is linked to allele $A$ of the donor parent. The genotype of the $F_1$ hybrid will be $Ab/aB$ and selection for allele $A$ in the first backcross generation will tend to pull along allele $b$, thus slowing the process of obtaining the desired $AB$ recombinant genotype. However, because allele $B$ is reintroduced with each backcross, there will be recurring opportunities for crossovers to occur. Assuming that selection is practiced only for $A$ in all generations, the allele being transferred, the probability of eliminating allele $b$ will be $1 - (1 - c)^N$, in which $c$ is the crossover rate between $A$ and $b$ and $n$ is the number of backcrosses to the recurrent parent. Hence if allele $B$ is located 50 or more crossover units from $A$, the probability that allele $b$ will be eliminated during six backcrosses is $1 - (0.5)^6 = 0.984$. In a selfing series with selection only for allele $A$, the corresponding probability is 0.50. These probabilities become progressively smaller, of course, with closer linkage between alleles $A$ and $b$; for example, if the recombination fraction is 0.01, the probability is $1 - (0.99)^6 = 0.06$, indicating that the unfortunate linkage would probably have been broken in about 6 among 100 backcross lines, in contrast to only about 2 among 100 lines derived by selfing. The probability of obtaining a desirable recombinant is clearly greater with backcross breeding than with selfing, assuming, as is often the case, that selection cannot be practiced against undesirable alleles that are linked to desirable alleles. However, should it be possible to practice effective selection against undesirable alleles that are linked to one or more desirable alleles under transfer, both selfing and back-crossing afford opportunities for obtaining desired recombinants. In fact, a selfing series may, under such circumstances, be more efficient than backcrossing because crossing over is possible in both male and female gametogenesis, whereas with backcrossing, effective crossing over cannot take place in the recurrent parent. Effective selection is, however, usually not possible with alleles affecting characters of low heritability. Hence, in dealing with characters of low heritability, the back-cross procedure is nearly always advantageous because it deals with all undesirable alleles automatically, regardless of heritability.

It is clear that recurrent backcrossing, even in the absence of selection, is a powerful tool for achieving homozygosity and that any population derived by backcrossing must rapidly converge on the genotype of the recurrent parent. When recurrent backcrossing is made the driving force in a plant-breeding program, the genotype of the recurrent parent will be modified only in respect to the allele or alleles being transferred, which must, of course, be maintained by selection. Prop-erly executed, backcross breeding programs thus allow all of the desirable charac-teristics of the recurrent parent to be recovered and retained, except for the possibility that alleles very tightly linked to the allele or alleles being transferred will be modified inadvertently. This is both the strength and weakness of the method. Recurrent backcrossing provides a precise and certain way of making gains of predictable value, with little possibility that uncontrolled segregation will produce unanticipated weaknesses that may be difficult to ascertain in necessarily finite periods of testing and evaluation of a novel backcross-derived variety. At the same time, the method has a disadvantage: it places an upper limit on the amount of advance, one that may be lower than that possible when segregation is less rigidly controlled.

## Selecting the Recurrent Parent

Because a series of backcrosses reproduces the genotype of the recurrent parent precisely, except for the character under transfer, it is obvious that backcross programs must be based on suitable recurrent parents. In well-established crops successful varieties, successful inbred lines, and $F_1$ hybrid varieties grown in any area almost certainly represent the end products of long periods of Darwinian evolution in which natural selection, selection by farmers and, more recently, efforts of plant breeders have each played a role. Many successful varieties in any ecological, agricultural, or societal situation have survived for long periods despite the release of numerous "improved" varieties intended as replacements. Enduring varieties and inbred lines are obviously the ones to be considered seriously for choice as recurrent parents in backcross breeding programs.

## Maintenance of the Character Under Transfer

In backcross breeding the level of heritability is rarely a problem of any consequence except for the character(s) under transfer, but for such characters high heritability is important. This is because selection must be practiced for the character(s) being transferred through each of several generations of backcrossing. At the same time all other characters will be taken care of automatically by the backcrossing procedure, with the result that the breeder need pay them little or no attention. Backcross breeding thus has its greatest ease of application when the character being transferred can be readily identified by visual inspection or by other quick and simple tests. As with all methods of breeding, characters governed by single alleles are easiest to manage. However, backcross breeding is by no means restricted to characters governed by alleles of one or even several genes. The number of alleles involved is often less important than the precision with which the character under transfer can be identified in segregating populations; a character of high heritability governed by several alleles might well be more easily transferred than a character of low heritability governed by only one allele. Regardless of the number of alleles governing the character to be transferred, it is essential that a worthwhile intensity of the character be maintained throughout a series of backcrosses. Some of the intensity of a character may be lost even when the control is primarily monogenic. This is often true in the transfer of disease-resistance alleles. Backcross-derived disease-resistance varieties carrying a "single" resistance allele may show some symptoms of disease when artificially inoculated, but may show no symptoms or, at the least, escape measurable damage under agricultural conditions. In some cases, even though there are some symptoms of disease, the causal organism has often been unable to maintain itself and the disease ceases to be a significant factor. Sooner or later, however, most pathogenic organisms "fight back" by developing novel virulent races. It has been possible in some cases to anticipate and circumvent this contingency by backcrossing additional potent resistance alleles into commercial varieties and deploying them if and when they are needed. This stratagem has proved effective in combating scald disease, caused by *Rhynchosporium secalis*, a highly resourceful foliar disease of barley characterized by numerous races governed by alleles of many different loci.

In dealing with quantitative characters, the donor parent may be chosen with the idea of sacrificing some of the intensity of the character to be transferred. Thus, an adapted short-grain variety of rice has sometimes been chosen as the recurrent parent and a long-grain parent as the donor parent in breeding medium-grain varieties of rice with the adaptedness and yielding ability of short-grain varieties.

## The Number of Backcrosses

If backcross breeding programs are to succeed with certainty, the complex, essential features of the genotype of the recurrent parent that endow it with superior overall worth must be recovered. Such recovery will almost certainly be primarily a function of the number of backcrosses, even though selection for the general type of the recurrent parent, especially in early backcross generations, can be effective in rapidly shifting the population toward the characteristics of that parent. Furthermore, the recurrent parent is frequently not composed of a single pure line but of a number of closely related pure lines that interact synergistically to give the recurrent parent its agricultural characteristics. Hence, enough plants of the recurrent parent must be involved in the backcrossing program to recapture the overall agricultural worth of the recurrent parent. Six backcrosses coupled with careful selection for the type of the recurrent parent, especially in early generations, has become more or less standard in backcross breeding programs in selfing species. It is commonly accepted that selection for the type of the recurrent parent, if based on moderate-size populations, is equivalent in effectiveness to one or two additional backcrosses without selection for type. After the third backcross, however, the population is likely to resemble the recurrent parent so closely that human-directed selection, on an individual plant basis, is largely ineffective except for intensity of the character being transferred.

Backcrosses can also serve a useful purpose in pedigree and bulk-population breeding programs. When, in a pedigree or bulk-population program, one of the parents is superior to the other parent in most of its characteristics, and especially if it is superior in overall worth, it is often advantageous to backcross the $F_1$ hybrid to the superior parent. One backcross will skew the segregating generations sharply toward that parent, thereby likely increasing the frequency of superior genotypes in late-segregating generations. Another alternative is to skew the segregation even more toward the superior parent by crossing the first backcross generation to the superior parent. Stopping the backcrossing early preserves the likelihood that transgressive segregation will contribute more than just a single superior desirable allele to the recurrent parent. However, a price attends this procedure, namely, relinquishment of control over the segregation that the backcross method regulates so precisely. As a consequence, the evaluation trials that are not necessary in standard backcross programs are necessary with programs that employ just "some" backcrossing. Programs that feature only "some" backcrossing are perhaps more accurately described as "backcross pedigree" or "backcross bulk" programs rather than as true backcross programs.

## Backcross Breeding Procedures

The general plan by which many backcross breeding programs have been carried out is as follows:

1. Selection of parents: The recurrent parent (parent $A$) is almost always a paramount variety of the area or an elite inbred line in outcrossing species, whereas the donor parent ($B$) usually possesses in high intensity some desirable, usually simply inherited, attribute in which parent $A$ is deficient.

2. The $F_1$ of the hybrid of $A \times B$ is backcrossed to parent $A$, giving a progeny segregating for the simply inherited desirable character of parent $B$. Several individuals with the desirable character of parent $B$ are selected and allowed to self-pollinate to produce $B_1F_2$ families, among and within which intensive selection is practiced for high intensity of the desirable character of parent $B$ and for the general features of the recurrent parent (parent $A$). The $B_1F_2$ plants are allowed to self-pollinate to produce $B_1F_3$ lines, among and within which further intense selection is practiced for high expression of the desirable trait of parent $B$ and for the general type of the recurrent parent (parent $A$).

3. The selected $B_1F_3$ plants are again backcrossed to parent $A$ to produce $B_2F_2$ ($A^3 \times B$) seeds. Plants from these seeds are backcrossed yet again to parent $A$ to produce third backcross ($A^4 \times B$) seeds.

4. The procedure of step 2 is repeated, with the fourth, fifth, and sixth backcrosses made in succession. An $F_2$ and $F_3$ are usually grown only after the sixth backcross, at which time intensive selection is practiced for the character being transferred, as well as for the general type of the recurrent parent.

5. A number of lines homozygous for the character being transferred, which are also as similar to the recurrent parent as possible, are selected, bulked, and the bulk increased and released for commercial production.

If appropriate numbers of plants are grown at each step in the foregoing program, the transfer of a single dominant allele can usually be accomplished through five backcrosses ($A^6 \times B$) with 53 plants from backcrossed seeds, 96 $F_2$ plants and 68 $F_3$ rows of 24 plants each. These numbers provide 0.999 probability of having at least one $Aa$ plant after each backcross and at least one homozygous $AA$ progeny in each $F_3$. In the case of species in which hand-pollinated artificial hybridizations are difficult (e.g., small-flowered legumes), the number of hybrid seeds required can be reduced by growing selfed $F_2$ and $F_3$ populations after each backcross and crossing on homozygotes. Somewhat smaller total populations are required to transfer incompletely dominant or recessive alleles, provided homozygotes can be recognized, making $F_3$ progenies unnecessary. Note that the program outlined previously does not call for successive backcrosses but for growing $F_2$ and $F_3$ generations only after the first, third, and sixth backcrosses. Note also that substantially larger numbers of plants are sometimes called for than the theoretical minima. The reasons for this are grounded in two considerations believed to outweigh the advantages of successive backcrossing and theoretically minimal population sizes. First, there is compelling evidence that selection for the types of the recurrent parent in $F_2$ or $F_3$ populations of modest size is equivalent to at least one, and perhaps even two, additional backcrosses in a continuous series without rigorous selection. Second, selfing after the first, third, and sixth backcrosses requires little effort and provides the breeder with populations sufficiently large to permit selection to be practiced for worthwhile intensity of expression of the character under transfer, as well as for the type of the recurrent parent.

## Improvement by Steps

Although the backcross method has its most straightforward application in the transfer of single monogenic characters, it also offers a number of procedures by which a variety can be improved for two or more monogenic characters more or less simultaneously, or by which additional alleles affecting one character can, if needed, be accumulated in one variety. Once a variety has been improved in one or more characteristics, the use of the latest backcross-derived individuals as recurrent parents will, of course, automatically preserve all previous improvements. This has frequently been the case in breeding for disease resistance in which resistance to some particularly damaging disease was first incorporated into a dominant variety, followed by later incorporation of resistance to less damaging disease using the earlier resistant genotypes as recurrent parents. In some cases the appearance of new disease races has made it necessary to add additional protective alleles. In other cases awns or other morphological characteristics were added because they had been found to improve yields and/or quality of product.

Another approach that has been used to improve a variety in two or more characteristics is to transfer the necessary alleles simultaneously in the same program. This procedure has the disadvantage that somewhat larger populations are required to transfer two alleles simultaneously than to transfer them separately. This disadvantage is even greater with three or more independent loci, because genetic complexity increases exponentially with the number of loci. Further, some characters may not be expressed dependably, thus delaying both or all transfers.

## Transferring Quantitative Characters

Backcross breeding is by no means restricted to transfers of traits governed by major genes. It has frequently been used to adjust the expression of quantitatively inherited characters, such as early flowering. In this case the usual procedure has been to cross a dominant variety or popular inbred line (the recurrent parent) with an early flowering stock (the donor parent), backcross to the recurrent parent, select and self-pollinate a number of $BC_1$ early plants that most closely resemble the recurrent parent, then select for earliness and for the plant type of the recurrent parent in the selfed $BC_1F_2$ family lines. Repeating this cycle one or more times will often provide early lines that adequately resemble the recurrent parent. In many cases it has been found that one or a very few alleles have major effects on earliness (and on other quantitative characters) and that the aforementioned procedure identifies these alleles and fixes them in individuals that are otherwise sufficiently like the recurrent parents. Thus important adjustments can often be made in quantitative characters without undue sacrifice in the essential properties of the recurrent parent.

## Applications to Cross-Pollinated Crops

Backcross improvement of outbreeding species differs in no fundamental way from that of self-pollinated species. However, with outcrossing species particular care is required to ensure that the sample of gametes taken from the recurrent parent accurately represents the allelic frequencies of that parent. The breeding of Caliverde alfalfa is a good example of the application of backcrossing to an out-crossing species. Mildew-resistant and leaf-spot–resistant plants found in California Common, the recurrent parent of Caliverde, were the source of resistance to these two diseases. Resistance of alfalfa to wilt disease was transferred from Turke-san, a winter-hardy type wholly unsuited for commercial production in California. About 200 plants of California Common were used to represent the recurrent parent in each of four backcrosses of resistant plants to the recurrent parent. The breeding of this variety, including seed increase for release to commercial growers, was completed in 7 years. Caliverde, aside from its resistance to mildew, leaf spot, and bacterial wilt diseases, was indistinguishable from California Common; it rapidly and completely replaced its recurrent parent in the commercial production of alfalfa in California.

## Influence of Environmental Conditions on Backcross Programs

One of the advantages of the backcross method, especially if the recurrent parent is a homozygous inbreeder, is that the program can be carried out in almost any environment that allows development of the character under transfer. With pedi-gree or bulk-population breeding programs the use of greenhouses or off-season nurseries is usually limited to tests for disease resistance or to adding generations, because efforts to evaluate agricultural performance are, in large part, wasted except under conditions closely approximating normal growing conditions of the result-ing variety. Backcross programs are, for the most part, not subject to this handicap because adequate backcrossing automatically restores the genotype of the recurrent parent. Hence, backcross programs can often be accelerated by growing additional generations in off-season nurseries, in greenhouses, or in cold frames. However, caution must be exercised if the recurrent parent is highly variable genetically and subject to rapid modification by selection, as may be the case with many varieties maintained by free outbreeding, such as alfalfa.

## USAGE OF BACKCROSSING IN PLANT BREEDING

Most of the examples discussed in this chapter have featured breeding for disease resistance. This is appropriate, because the backcross method has been used for this purpose perhaps more than for any other single purpose. Moreover, many of the early descriptions of backcross breeding in the plant breeding literature were based on transfers of dominant monogenic resistance alleles into highly homozygous cereal varieties. It should be emphasized, nevertheless, that the backcross method is suited for numerous other purposes as well, including transfers of recessive alleles and adjustment of any trait under transfer that is at least moderately heritable. Although the early literature is replete with successful applications of the method, more recent applications have usually not been publicized. One reason is that the method is so simple and straightforward that many breeders have regarded an-nouncement of results as superfluous. Another, unjustified, reason is that some breeders have thought that publicizing the results of such a simple, straightforward, and predictable method might expose them to being labeled "hack" breeders.

The main reason the backcross method has come to be particularly associated with breeding for disease resistance is probably related to the devastating effects that rusts can have on the productivity and quality of wheat, the "staff of life." The place of North American hard-red-winter and hard-red-spring wheats in world markets depends on their abundance and noted baking quality. The appearance of new and devastating races of rust that occurred with disturbing regularity in the 1940s was a matter of international concern, especially the appearance of the exceptionally virulent race, 15B, of stem rust. New high-yielding varieties of wheat resistant to these races had to be developed quickly, and programs of backcross breeding that avoided the necessity of lengthy evaluation trials to recover important quality characteristics were widely adopted for this purpose. The danger of losses owing to new physiological races of rust was undoubtedly the main factor that induced many North American wheat breeders to adopt backcross programs, which succeeded spectacularly. The method was also used widely in South America and Australia. In Europe, Asia, and Africa, however, where cereal diseases are generally less destructive, the backcross method has been used considerably less frequently. Nevertheless, the value of the backcross method for handling a wide variety of problems in plant breeding has been amply demonstrated, and the use of the method has continued, but nearly always with little fanfare

Another aspect of backcross breeding has been described by Borlaug (1957) as follows:

> There are many undesirable aspects associated with precipitous changes in varieties. Farmers are reluctant to shift from a proven variety to a new unknown variety. Their reluctance is based on familiarity with the old variety which permits them to exploit to a maximum its potential yielding ability. They know the best rates and dates of seeding of the old variety for their local conditions. When a new variety is introduced these considerations must again be worked out by the grower for his own local conditions before he is able to utilize the new variety in such a way as to approach its potential optimum productivity. . . . The conventional backcross method of plant breeding comes closest to overcoming this problem. It provides, if properly carried out, new varieties which are for all practical purposes phenotypically identical to the recurrent parent and therefore readily received by farmers, processors and consumers.

Many breeders have not followed the conventional backcross process to reconstitute the recurrent parent precisely, but have used only one, two, or three back-crosses to retain the possibility of segregation for agronomic or horticultural characteristics such as yield and adaptedness. The possible benefits of compromises of this sort are enticing, but there are serious possible disadvantages. Time has been and remains a particularly important factor in the production of new varieties; it usually takes as long or longer to evaluate properly the potential of a new variety than it takes to produce and to bring such a variety to the point at which intensive evaluation can begin. Moreover, evaluation in one place can rarely be assumed to be accurate except, possibly, for that particular place. As a result of the pressures of time and expense, many "promising" genotypes have been released to commercial production prematurely. In addition, the success or attainment of eminence of breeders, and the institutions they represent, has sometimes been measured by the total numbers of varieties that breeders have released. This is certainly a much less appropriate measure of accomplishment and achievement than the number and the extent of success of the truly worthy varieties that breeders have released to commercial production. The success of the backcross method has been amply demonstrated in situations of emergency. In such situations it appears both reasonable and prudent for breeders to put first things first, that is, to confront any potential emergency directly and without delay, relegating less pressing matters to secondary positions of priority. This leaves open the question as to when less certain breeding methods that depend on transgressive segregation should be instituted. An issue such as this clearly depends on how long it is likely to take to develop a variety capable of superseding a backcross-derived emergency variety in overall worth. It is possible, but unlikely, that this might happen soon enough that a backcross-derived emergency variety need never be released; however, should this happen, little wastage will likely have occurred because backcross programs require minimal expenditures of time, labor, and other resources—and even inexperienced breeders rarely fail to produce a worthwhile product.

# FOURTEEN

# *Breeding Hybrid Varieties of Outcrossing Plants*

Many species of cross-pollinated cultivated plants are seed-propagated, whereas other outcrossing species propagate themselves vegetatively as clones in nature and are also propagated as clones in cultivation. Breeding procedures appropriate to exclusively seed-propagated outcrossing crop species are the concern of this chapter; breeding procedures appropriate to species that are propagated in cultivation either exclusively or largely by vegetative means will be considered in Chapter 15.

In a population of seed-propagated randomly mating cross-pollinated plants every individual can be expected to be homozygous at many loci, but also heterozygous at many loci. As a consequence of the segregation and recombination that follows open outcrossing among genetically different plants, alleles of many loci are reshuffled and regrouped into vast numbers of multilocus allelic configurations each generation; an allele that is homozygous in one generation may become heterozygous in the next generation and vice versa. Analogously, alleles that are part of one multilocus configuration in one generation may be found in different multilocus combinations of alleles in subsequent generations. When many genetic loci are represented by more than one allele, it is likely that almost limitless numbers of genetic combinations will be present, and it is hence unlikely that any two individuals with precisely the same multilocus genotype will occur in successive generations in any large open-pollinated population. Under natural selection, whether in nature or in cultivation, only those individual alleles and those multilocus combinations of alleles that promote adaptedness and reproductive fitness in the particular shifting sets of environmental circumstances to which the population has been exposed are likely to have been favored and to have increased in frequency at the expense of alleles or combinations of alleles less favored in those particular sets of environments. In cultivation the shift toward better-adapted genotypes in the local environment can be accelerated by effective natural, farmer- and/or breeder-directed selection. Nevertheless, increase in yield is likely to be very slow even under intense selection, such as the selection that was practiced with U.S. corn in the open-pollinated period from about 1860 to 1930 (Figure 5-1). Trueness to type in an open-pollinated population is a statistical feature of the population as a whole; it is not a characteristic of individual plants, which is often the case with

pure-line populations. Hence, the constantly changing composition of open-pollinated populations virtually guarantees that such populations will be less efficient
and lower yielding than monogenotypic populations composed solely of numerous
identical copies of truly exceptional genotypes that have been identified as superior
through repeated testing by nature, by farmers, and/or by breeders. It is, consequently, not surprising that a common goal in the breeding of seed-propagated
outcrossing species has been the development of monogenotypic single-cross
hybrid varieties. Single-cross hybrid varieties are composed of numerous identical
copies of a genotype that has been identified by means of extensive testing programs
as excelling in adaptedness, in yielding ability, and, likely, in product quality as well.
Because the two inbred parents of such hybrids are homozygous at all loci, all
single-cross $F_1$ hybrid plants, even though they carry different alleles at many loci,
have the advantage of being homogeneous for the same superior highly selected
multilocus genotype. Such varieties also have the advantage, which they share with
inbred line varieties of selfers and with vegetatively propagated single-clone varieties, of being precisely reproducible year after year. However, seed-propagated $F_1$
hybrid varieties have a disadvantage, namely, that on reproducing sexually they
segregate in an unmanageable manner so that very large numbers of different
genotypes appear in the $F_2$ generation and very few, if any, among these segregants
perform as well as the single genotype of the $F_1$. As a consequence, yields of
seed-propagated $F_2$ populations resulting from segregation during reproduction in
the $F_1$ are nearly always at least 10%, and often more than 30%, lower in yielding
ability than their parental monogenotypic $F_1$ hybrid parents. Consequently, growers must obtain a new supply of $F_1$ seed each generation. In high-input agricultures
seed is inexpensive as compared with the gains attending the growing of hybrid
varieties; hence, this disadvantage is relatively minor. Although some plant breeders
regard the breeding of single-cross hybrids as only a highly specialized form of
traditional open-pollinated breeding methods, most breeders consider single-cross
methods to be such a great and strikingly successful departure from all previous
methods as to justify classifying this method as a new and intrinsically different
breeding procedure. The earliest ideas basic to breeding hybrid varieties date back
to the 1870s and 1880s (Beal 1876–1882) and the first decade of the 1900s (Shull
1908, 1909). We turn now to a remarkable series of complexly intertwined developments that occurred in the first 50 years or more of the 1900s, developments
featuring repeated cycles of population improvement within originally open-
pollinated source populations with the goal of enhancing the value of the inbreds
that could be extracted from the improved source populations. Thereafter test
crosses to determine general combining ability (GCA) and/or specific combining
ability (SCA) were added as crucial tests needed in the breeding of the superior
inbreds to identify the best monogenotypic $F_1$ hybrid varieties. Breeding procedures
appropriate to seed-propagated outcrossing crop species, such as corn, will be
described in this chapter, whereas procedures appropriate to producing superior
clonal varieties and/or varieties that are propagated from planting mixtures (including mixtures of clones and/or various kinds of seed mixtures), will be described
in Chapter 15. Before attempting any of the above tasks, however, it is appropriate
to explore the development of the ideas on which population-improvement methods in outcrossing crops are based. It is fitting that this be done largely in terms of

studies conducted with corn (*Zea mays*). No other species can be selfed and crossed so reliably or as inexpensively, and, as a result, it has been possible to investigate breeding methods in corn in ways and with precision unrivaled with any other outcrossing species.

No discourse concerning the development of methods of breeding hybrid varieties could be entire without reference to two nongenetic features that were essential to the rapidity with which the procedures basic to breeding hybrid varieties were ultimately adopted. The first feature has to do with plant breeders' rights. Breeding homozygous lines, open-pollinated varieties, and clonally reproduced varieties have, in general, not been rewarding for commercial breeders in the absence of legally established plant breeders' rights. Once released, such varieties can be propagated and sold by anyone without profit to the breeder. Hybrid varieties, in contrast, offer built-in economic protection; the breeder can retain control of the parental inbreds and sell only hybrid seed. This is because the next and later generations produced from $F_1$ hybrids yield much less than the $F_1$ hybrids; hence growers of hybrid varieties must, of necessity, return each year for $F_1$ hybrid seed. Although hybrid varieties were first developed by publicly supported institutions, private breeding firms soon produced hybrids equal or superior to those of state-supported experiment stations, and hybrid corn rapidly became an eminently successful commercial operation. The resources put into producing hybrid corn by industry were quite substantial as compared with those available to state-supported plant breeding; moreover, seed of truly excellent hybrid varieties was being sold inexpensively. An additional feature was that hybrid corn was a major component in the passage from traditional agriculture to intensive technology-based agriculture. Many commercial hybrids were soon bred specifically for adaptedness to increasingly higher plant densities and increased applications of fertilizers. Hence, hybrid corn varieties were soon developed that were highly adapted wherever agricultural technologies were sufficiently advanced to benefit economically, and at present hybrid variety breeding methods are increasingly being extended to other crops (Chapter 16).

## POPULATION IMPROVEMENT

Population-improvement methods fall naturally into two categories: those based on phenotypic selection without progeny testing and those based on phenotypic selection with progeny testing. Traditionally, the practice of early farmers and early corn breeders was to select numerous plants with particular ear, kernel, or other characters they believed to be favorable, and thereafter, generation after generation, they grew the next generation, or generations, from mixtures of seeds produced on selected plants of the previous generation. Such mass selection without progeny testing is the original and simplest breeding method, and, historically, it has been the most widely used method for breeding all seed-propagated crops, including both inbreeders and outbreeders. Selection based on visual evaluation, or on simple measurements of the phenotypic characteristics of seed parents of individual plants, is referred to as phenotypic mass selection. This type of mass selection was conceptually simple and easy to conduct; further, its single-generation breeding cycle was appealing. However, a major weakness of such mass selection, not fully appreciated until late in the nineteenth century, was that it did not allow satisfactory control of pollen parents in outcrossing species. Under open-pollination the successful pollen could come from any random plant in the population, including plants less desirable phenotypically than the selected seed parents. Another weakness, also not widely recognized until late in the nineteenth century, was that phenotypic mass selection depends for success on high heritability—that is, on high regression of offspring on parents. One of the late-nineteenth-century variants of phenotypic mass selection, practiced specifically in corn, was ear-to-row breeding in which ears from selected plants were planted in different rows and only those progeny rows that were judged visually to be superior were harvested for further selection. This progeny-testing procedure was effective, but only for very highly heritable characters such as oil content. In practice, ear-to-row selection turned out to be largely ineffective, or at best weakly effective, for characters, such as yield, that are considerably affected by environment. A further later refinement sometimes included in ear-to-row selection was the selection of not only female parents but also superior male parents from the same population. The selected male parents were then crossed with the selected female parents, with the expectation that the progeny obtained would be superior to progeny that resulted from outcrossing with random male plants in the population. However, the small additional gains sometimes realized rarely justified the added labor requirement for the improved control of parentage and/or the added time and labor required for progeny tests. Thus, ear-to-row breeding, even with embellishments, came to be recognized in the early 1900s as, at best, only a marginally effective method of population improvement. It was not until many years later that improved yield-testing techniques and improved ways of estimating heritabilities came into wide use.

## HYBRID VARIETIES

Late in the nineteenth century and early in the twentieth century it was becoming increasingly apparent to corn breeders and farmers alike that traditional breeding methods (repeated mass selection for particular plant characteristics, especially selection for ear conformation, such as was eagerly promoted in corn shows), as well as the newer ear-to-row methods, were not producing satisfactory results in the form of improved yields (see Figure 5-1). The first hint of a way to solve the problem came with the reports by Beal (1876–1882) that hybrids between different varieties of corn intervarietal hybrids, were usually more productive than open-pollinated parental varieties. Although Beal did not provide detailed yield data, he asserted that the yields of intervarietal hybrids of corn were higher than those of their parents, sometimes by as much as 40%. But neither breeders nor farmers of the time were ready for such a considerable departure from established ways of corn breeding, with the result that Beal's underlying idea, to exploit "hybrid vigor" using controlled intervarietal crosses, was not accepted until more precise data on the effects of inbreeding, and the often favorable effects of crossing selected inbreds, became available as a result of the independent work of East (1908) and, particularly, of Shull (1909). The more important effects of continued self-fertilization reported by East and Shull can be summarized as follows: (1) A large number of lethal and subvital types appeared in early generations of selfing; (2) the progeny of original parental plants rapidly separated into distinctive lines that became, with continued inbreeding, increasingly uniform for various morphological and functional characteristics such as height, ear length, and maturity; (3) all inbred lines decreased in vigor and fecundity, until many could no longer be maintained even under the most favorable cultural conditions; and (4) all lines that survived showed a general decline in size and vigor; but (5), and by far most important, many unrelated inbred lines when intercrossed gave superior $F_1$ hybrids.

From the earliest recorded history, inbreeding has been associated with lessened vigor, reduced fecundity, and the appearance of increased numbers of defective off-types in inbred populations. Among the consequences were pronouncements in many human societies that prohibited marriages between relatives of specified closeness (forbidden degrees) of kinship. Inbreeding has not, however, consistently been regarded as something to be avoided. For example, consanguineous marriages have been favored in some human societies, sometimes for the purpose of preventing "dilution of superior blood lines," but also for the nonbiological purpose of maintaining property and/or power within families.

The development of many modern breeds of livestock provides another example of the potential biological usefulness of inbreeding. Starting about 1700 it had become common practice for livestock breeders to mate the offspring of outstanding males with their offspring in particular lines of descent to preserve or to fix desirable characteristics and to eliminate undesirable characteristics. However, such "line-bred" stocks sooner or later "ran out," and it became necessary to cross outside the line to restore fecundity and productivity. The system thus became one of inbreeding (breeding within outstanding lines) to concentrate desirable qualities, coupled intermittently with crossbreeding to prevent degeneration. Darwin (1868), who described this process in detail, concluded that "although free crossing is a danger on one side which everyone can see, too close inbreeding is a hidden danger on the other." Scientific information concerning the effects of outcrossing in plants dates from the experiments of the plant hybridists of the eighteenth century. Kolreuter is reported to have noted that hybrids are often remarkably luxuriant, whereas Sprengel is reported to have reached the conclusion from studies of the relationships between flowers and insects that nature usually "intended" that flowers should not be pollinated by their own pollen. Darwin (1876) recorded the results of his own observations and experiments; he reached the conclusion that self-fertilization is an unnatural and harmful process. Apparently, the first inbreeding experiments with corn were those of Darwin. Unfortunately his experiments with corn were continued for only a single generation and hence were not very informative. Darwin did, however, observe the dramatic negative effects of even a single generation of selfing and concluded that "nature abhors self-fertilization."

Shull was also impressed by the obvious deleterious effects of inbreeding corn, especially the large numbers of lethals brought to light in the early generations of inbreeding, as well as the decrease in vigor and fecundity of longer-surviving lines. By 1908, however, after many years of observations, he wrote that a field of open-pollinated corn "contains" many complex hybrids and that corn breeders should strive to maintain the best "hybrid combination." Earlier he had accepted the concept held by some corn breeders that selfing has "deleterious effects," but Shull did not recognize until 1907 or 1908 that crossing often had advantageous effects. When Shull began to make crosses among his inbred lines, he found that the degree to which vigor and productivity were restored on crossing were functions of the origin of the lines crossed. Crosses between sibs within a self-fertilized line showed little improvement over self-fertilization in the same family; however, the difference between selfing and sib-crossing became progressively greater as the numbers of generations of inbreeding increased. Still more important, when inbreds derived from different original open-pollinated plants were crossed at random, the average response was a return to the vigor and productivity of the original population before inbreeding had commenced. Some hybrid combinations showed greater improvement than the average, whereas others showed less improvement. But Shull emphasized that there was no overlapping of the hybrids as a group and the inbred parents as a group; the hybrids in all cases were superior to the inbred parents. By 1908 these experimental results forced Shull to a conclusion that had been unappreciated over the centuries: that it was crossing that had advantageous effects and that the breeding of corn must have as its objective the recovery of occasional, infrequent, but vigorous and productive plants, presumably rare fortuitous hybrid combinations, such as he had observed occasionally in his open-pollinated populations. Shull's insights and his experiments in the next few years adumbrated many of the intricate and complexly interrelated procedures that are used at present in breeding hybrid cultivars of corn and other crop species, as well as in breeding many domesticated animal species. However, before proceeding to discussion of hybrid varieties, it seems appropriate to introduce two opposing but widely accepted hypotheses advanced in the early decades of the 1900s to explain the apparently conversely related phenomena of "inbreeding depression" and "hybrid vigor" or "heterosis," the latter a term coined by Shull to express the advantageous effects of hybridity.

## THE DOMINANCE AND OVERDOMINANCE HYPOTHESES

The overdominance hypothesis, proposed independently by East and by Shull in 1908 and 1909, postulated a physiological stimulus to development that increases with the diversity of the uniting alleles of the same locus in heterozygotes—that is, that many single loci exist at which the phenotype of heterozygotes is superior to that of corresponding homozygotes. In Mendelian terms this implies that there are single genetic loci at which particular heterozygotes are superior to corresponding homozygotes and that such stimulus increases with greater differences between the interacting alleles in heterozygotes. This idea has been variously designated as single-gene heterosis, superdominance, or stimulation of divergent alleles; however, the term usually applied to the concept has been *overdominance*. East elaborated this idea further in 1936 when he proposed a series of alleles at individual loci $a1$, $a2$, $a3$, $a4$, . . . of gradually increasing divergence in physiological function; thus, an $a1a2$ heterozygote was postulated to be less vigorous than $a1a3$, $< a1a4$, and so forth. At the time East formulated this hypothesis there was no direct evidence that any heterozygote lay outside the range of the corresponding homozygotes. Single-locus "heterosis" ($Aa > AA, aa$) has subsequently been inferred from measurements of seed size in heterozygotes versus homozygotes at some known major loci (e.g., chlorophyll mutants in barley) and from estimates of the fitnesses of heterozygotes relative to corresponding homozygotes in populations of corn, barley, beans, and other species. Others argued, however, that such apparent overdominance might be due to closely linked loci, an issue that is difficult to resolve experimentally. Regardless, clear-cut cases of single-locus heterosis have not appeared to this day, and the absence of such cases has been a deterrent to general acceptance of the overdominance hypothesis.

Another perhaps equally widely accepted genetic hypothesis to explain "inbreeding depression" and the apparent conversely related phenomenon of "hybrid vigor" also had its beginnings soon after the rediscovery of Mendel's work. This hypothesis, proposed by Davenport (1908) and Bruce (1910), started with the widely accepted postulate of the time that cross-fertilizing populations are made up of large numbers of genetically different individuals, many carrying deleterious recessive alleles concealed in heterozygotes. When such individuals are inbred, there is perforce an increase in homozygosity and various morbid homozygous recessive types appear, including plants that have defective flowers, defective seeds, lack of chlorophyll, and the like. These types are unable to survive or to reproduce effectively and so are quickly eliminated from inbred populations. Other characters also came to light that do not necessarily lead to immediate extinction of the afflicted individuals in inbred populations (e.g., dwarfness and partial chlorophyll defectiveness), but nevertheless impose serious handicaps. The appearance of such types on selfing appeared to provide an explanation for part of the injurious effects of inbreeding. Segregation occurs at a decreasing rate within lines as inbreeding progresses and sooner or later brings about near homozygosity, ultimately resulting in lines that carry different alleles and different complexes of alleles of different loci. Some lines by chance apparently receive more favorable and fewer unfavorable alleles than others, accounting for the differences observed in degree of inbreeding depression in different lines. Inbreeding depression, according to this hypothesis, is therefore not intrinsically a process of degeneration but a straightforward consequence of Mendelian segregation. Thus, the injurious effects of inbreeding are not produced by the process of inbreeding itself, as believed by many early biologists and philosophers, but the magnitude of such effects are directly related to the number and kinds of Mendelian "factors" in the original population. Under this hypothesis the intercrossing of inbred lines should sometimes lead to the formation of hybrids in which deleterious alleles, usually recessives contributed by one parent, are immediately hidden, as they were in the original open-pollinated stock by alleles contributed by the other parent. The precise degree of response to crossing should therefore be a function of the genotypes of particular inbreds. Some genotypes, when crossed, should complement each other nicely to produce hybrids better than the average of the original open-pollinated variety, whereas others might not "nick" well owing to the particular unfortunate combinations of alleles they happened to receive by chance during the haphazard segregation that occurred during the inbreeding process.

Objections to the dominance hypothesis were quickly raised on two grounds. First, if the hypothesis were correct, it should be possible to obtain inbred lines homozygous for all favorable dominant alleles. Such lines should be like the $F_1$ in vigor, and they should be true-breeding. However, high-yielding inbred lines were not found in very extensive inbreeding experiments. Jones (1917) reconciled this apparent discrepancy by pointing out that many Mendelian alleles affect growth and that each chromosome would be expected to carry many such alleles. In addition, any single chromosome would be expected to carry some favorable dominants and some unfavorable recessives. A series of precisely placed crossovers and fortunate segregation distributions would, therefore, be required to obtain all favorable alleles in one gamete, a process that would almost certainly require very large numbers of sexual cycles to achieve. Thus, Jones was apparently the first to recognize the important point that short chromosome blocks, not single alleles, are the units of genetic transmission in sexual reproduction. The second objection to the dominance hypothesis was directed at the apparently symmetrical distributions that were observed for heterotic phenotypes during inbreeding. If heterosis were due solely to dominance of independent genetic factors, distribution curves for such characters should be skew rather than symmetrical, because dominants and recessives would be expected to be distributed according to expansion of the binomial $(3/4 + 1/4)^N$. Jones reconciled the symmetrical distributions actually observed with the dominance hypothesis on the basis that linkage between favorable and unfavorable alleles within chromosome blocks should be counterbalancing and thus should lead to symmetrical distributions. However, Collins (1921) pointed out that even in the absence of linkage within chromosome blocks, skewness would be difficult to detect if many loci were involved and, moreover, that the chances of recovering any completely homozygous types would be small so long as the numbers of genes involved were at all large. These several ideas came to be called the *dominance* or *dominance of linked genes* hypothesis.

## POPULATION STRUCTURE AND HETEROSIS

Inbreeding is regularly deleterious in many species (e.g., alfalfa, corn, drosophila) and sometimes deleterious in other species (onions, sunflowers), but it has little or very little effect in still other species (cucurbits and the self-pollinated species). Many investigators have postulated that these differences depend on mating systems and the influence they have on the kind, the amount, and the ways in which genetic variability is organized in populations. In species that reproduce by self-fertilization, homozygosity is normal; deleterious mutants become homozygous soon after their origin and are eliminated promptly. Accordingly, populations of such species ultimately become adapted to homozygosity and develop a genetic organization that Mather (1943) called "homozygous balance." The type of population structure characteristic of self-pollinated species also appears to exist in some cross-pollinated species that have very small effective population sizes. Thus, a single farmer may require only a few plants to satisfy his needs, so that population size is restricted generation after generation to very few interbreeding individuals. Under such circumstances most loci would be expected to soon become and remain homozygous, thus developing so-called homozygous balance. This may explain why some outcrossing species such as sunflowers and cucurbits show little inbreeding depression.

Populations that suffer severe inbreeding depression have consistently been found to carry large numbers of recessive alleles sheltered in heterozygotes. In *Drosophila pseudoobscura*, for example, every fly among the many thousands that have been tested has been found to harbor at least one recessive lethal, and lethals are by no means the only component of the load of unfavorable alleles carried by such species. The load of deleterious mutations in alfalfa appears to be at least as great as that in various *Drosophila* species, and the load in corn is perhaps nearly as great. However, many investigators doubt whether the heterotic responses observed in such species can be accounted for by the dominance hypothesis. One reason for skepticism is that there is little convincing evidence that such commonly observed mutants have much effect on yield. Wentz and Goodsell (1929), for example, compared the frequencies of defective seed, seedling, and mature-plant recessives in 19 open-pollinated corn varieties with the yields of these 19 varieties and detected no relationship. Woodward (1931) self-pollinated a large number of vigorous plants of three open-pollinated varieties of corn and classified the resulting selfed lines for defective mutants. Seeds from selfed ears free from defective mutants were then composited to reconstitute the three varieties. The yields of the "purged stocks" were compared with the yields of the original stocks and found to be nearly exactly the same.

Crow (1984) calculated the maximum heterosis possible under the dominance hypothesis and concluded, "It seems probable that (the dominance hypothesis) may explain a major part of the loss in vigor with close inbreeding of random mating strains and the recovery of vigor on crossing. . . . However, it cannot account for increase in vigor following the crossing of artificially inbred strains much beyond the level of the equilibrium population from which the inbred lines were derived."

## THE NATURE OF HETEROTIC LOCI

Increasing evidence that the dominance hypothesis, at least in its simplest form, was inadequate to account for the heterosis actually observed in various species led to increased interest in the nature of "heterotic loci," their relationship to the selective advantage of heterozygotes, the production of balanced phenotypes, and the development of *homeostatic* systems (Lerner 1954a, 1954b; Dobzhansky and Wallace 1953). The earliest view concerning the nature of heterotic loci was formulated by East and Shull in 1908, as we saw earlier, in physiological language. Shull's concept was that "hybridity itself—the union of unlike elements, the state of being heterozygous—has a stimulating effect on the physiological activities of the organism." This idea, restated in genetic terminology by East in 1936, attaches to heterozygotes a versatility of development that is expressed in terms of alternative pathways in metabolism. These presumably arise in biochemical versatility stemming from existence of different alleles at the same locus. Because such versatility is not open to homozygotes, it must be associated with particular loci, which can be called heterotic loci. A necessary corollary is that heterozygosity is essential for vigor. The "homeostatic system" hypothesis (Lerner 1954a) attached no physiological virtue to the heterozygous state in itself; it reversed the causal sequence and made the physiological virtues observed in many heterozygotes consequences of selection under an outbreeding system. According to this second hypothesis "relationally balanced" (Mather 1943) or "coadapted chromosomes" (Dobzhansky and Wallace 1953) are produced as a result of long-term evolutionary development of balanced complexes of genes that have been integrated into the population gene pool.

Precisely what a heterotic locus represents in terms of physical structure remained a matter of spirited debate. Studies of pseudoallelism (e.g., Green and Green 1949; Lewis 1951) suggested that the unit of specific physiological interaction is probably not a single Mendelian locus in the classical sense but a complex of more than one, and perhaps several to many, tightly linked genes. This had been implied by Jones (1917) when he formulated his hypothesis of dominant linked factors: if linkages are close, crossing over between loci will necessarily be rare, with the result that small chromosome blocks rather than individual Mendelian loci act, at least in the short term, as units of hereditary transmission. This proposition has appeal because chromosome blocks vary in size with the closeness of linkage and the number of sexual generations, thus providing opportunities for study of evolutionary change within small but potentially complex hereditary units. We now turn to the much studied effects of inbreeding in corn as an approach to understanding both inbreeding depression and combining ability.

## INBREEDING DEPRESSION

One of the most obvious features of large outbreeding populations, including large populations of corn, is that intensive inbreeding causes a veritable rain of apparently single-locus defectives, usually recessives, to appear in the early generations of inbreeding. However, numerous low-vigor types, including types with phenotypes that suggest control by numerous loci with relatively small cumulative effects, also appear. Furthermore, there is a decline in general vigor and in fertility so that with continued inbreeding many lines become increasingly weaker and some lines ultimately die out. Such inbreeding depression is accompanied by steadily increasing uniformity within lines as homozygosity increases. This suggests that the numbers of loci (short chromosome segments) that are responsible for inbreeding depression, and hence also for heterosis, must be large, almost certainly much larger than the numbers of chromosome arms. With so many loci involved, complex interactions among loci seem inescapable. Moreover, there is increasing evidence that many loci carry out their functions partly, or even principally, through epistatic interactions and that the more subtle traits (e.g., such traits as yielding ability and adaptedness to specific facets of fine-grained environments) are frequently attributable to complex interactions among several to many loci (Chapters 5, 10, 11). The effects of various genetic processes such as those discussed earlier (e.g., dominance, inbreeding, linkage, epistasis) on population structure in corn presumably are expressed progressively over several generations of selfing until more or less full homozygosity is approached in generations $F_6$ to $F_7$ of selfing. In selfing generations ($F_2$ to $F_5$) the original chromosome structure of the base open-pollinated populations undergoes rapid and intricate changes generation after generation (there are perhaps two or three chiasmata per chromosome per generation) as segregation and recombination break up and form new chromosome blocks, each probably including several to many loci. Hence, linkage, dominance, increasing homozygosity, and epistasis interact in producing highly complex changes in gene organization and expression.

It seems likely that attempts to unravel the rapid and highly complex changes that occur during periods of inbreeding itself must nearly always fail. Such attempts have, in fact, defied genetic interpretations other than the very simplest, that of attributing inbreeding depression simply to fixation of numerous unfavorable recessives and attributing heterosis to the converse effect, namely, the covering of effects of unfavorable recessives by corresponding dominant alleles fixed in other inbred lines. If it were as simple as that, it might be possible to fix heterosis in true breeding lines. However, even though various types of "population improvement" programs have succeeded in increasing the performance of inbreds substantially during the last half century or more, the performance of the best inbreds still falls quite short of that of good hybrids. This raises a question: Is inability to develop high-yielding homozygous inbred lines merely a numerical problem stemming from the necessity of fixing very large numbers of favorable alleles in inbreds, or must other genetic factors also be taken into account? True overdominance is almost certainly rare; however, dominance and/or epistasis, together with linkage, are known to be capable of simulating overdominance, and so all such factors must be taken into account. Hence, studies of the combining ability of different fixed inbred lines may provide a less complex way to estimate the relative importance of the various genetic processes that affect heterosis than studies of its apparent inverse, which is inbreeding depression.

## COMBINING ABILITY

Starting with a hypothetical open-pollinated base population in equilibrium, postulating $p = q = 1/2$ for two alleles, designated $A$ and $a$, for each of many different loci, assuming relative fitness values $AA = Aa > 2$ $aa = 1$ (full dominance of $A$) for each locus, and assuming no interactions among loci, the relative mean fitness of the base population is expected to be $1/4 AA$ (2) $+ 1/2 Aa$ (2) $+ 1/4$ $aa$ (1) $= 0.5 + 1.00 + 0.25$, respectively, for total fitness $= 1.75$. In actual populations dominant $A$ may, in some or even in many cases, be favored over recessive $a$; hence, the $p$ actually realized during inbreeding may be $> q$ and total population fitness may be $> 1.75$. In addition, if some heterozygotes $Aa$ exceed $AA$ for some loci (overdominance), this also would tend to increase the mean population fitness of the equilibrium population above 1.75. However, under selfing, and with so many loci, no inbred line could escape possibly substantial fixation for very long, leading to severe inbreeding depression within all lines as well as in the inbred population as a whole. Moreover, should any favorable epistatic complexes exist in the population, the random segregation and recombination associated with inbreeding would almost certainly quickly break up such complexes of alleles, especially favorable combinations of alleles of loci located on different chromosomes or at some distance apart on the same chromosome.

This has, in fact, proved to be the case. For example, in corn 5 to 7 generations of selfing leads to essentially complete homozygosity featuring ~1/2 $AA$: ~1/2 $aa$ genotypes, so that the mean fitness of the fully inbred population is expected to be < 1.0, that is, less than half that of the original equilibrium population (~1.75), which is, in fact, usually about the level of fitness observed in $F_5$ to $F_7$ populations in which minimal selection has been practiced during the inbreeding process. The precise genetic composition of such inbred populations is expected to be extraordinarily difficult to determine, especially when the extensive segregation and recombination that occurs in generations $F_2$ to $F_4$ of inbreeding are taken into account. Many thousands of novel genotypes are likely to be formed in these generations. One thing seems certain, however, namely, that randomly chosen pairs among the myriad of fully or near-homozygous lines that emerge from these chaotic processes in generations $F_4$ to $F_6$ could only rarely be expected to have received those particular combinations of alleles $A$ and $a$ and other alleles that complement one another to give performance equal or superior to that of the base random-mating population.

If, as is commonly supposed, alleles $A$ and $B$ of most loci are fully or nearly fully dominant to alleles $a$ and $b$, the four hybrids that can be produced by inbred lines carrying the two possible homozygotes ($AA,aa$ and $BB,bb$) will each occur in frequency $1AA:2Aa:1aa$. Hence, with full or near-full dominance, 3/4 of such combinations (e.g., $AA + 2Aa$ with $BB + 2Bb$) will be nearly equally favorable, but with overdominance only 2/4 of combinations ($2Aa$ and $2Bb$) will be highly favorable. With $N$ loci the best possible inbred $\times$ inbred pairs are thus expected in frequencies of about $(3/4)^N$ and $(1/4)^N$, with full dominance and overdominance, respectively. Furthermore, the frequencies of the best-possible inbred pairs are expected to decrease rapidly with increases in the number of loci; hence, if 10, 20, and 50 loci are involved, the expected frequencies $(3/4)^{10}$, $(3/4)^{20}$, and $(3/4)^{50}$ decrease to 0.056, 0.003, and ~ $6\times10^{-7}$ with dominance and to $(1/2)^{10}$, $(1/2)^{20}$, and $(1/2)^{50} = 0.001$, ~$9.5 \times 10^{-7}$ and ~$9 \times 10^{-16}$ with overdominance, respectively. Pseudo-overdominance (epistasis) with some recombination would be expected to produce intermediate values. Hence the average random inbred $\times$ inbred cross is expected to involve only very small numbers of favorably interacting alleles on average, and random inbred $\times$ inbred crosses are expected to perform, on average, at approximately the same level as in the original source populations. Such poor performance is, in fact, in accord with the performances actually observed when large numbers of inbred lines derived from the same source population are pair-crossed at random. This, in turn, suggests that full or near-full dominance is the usual interactive situation in respect to inbreeding depression and heterosis. Moreover, outstandingly good $\times$ outstandingly good, as well as outstandingly poor inbred $\times$ outstandingly poor inbred combinations are both expected to be very rare. All three of these expectations are understandable (1) if most allelic interactions are in the full or near-full dominance range, (2) if true overdominance is rare, and (3) if pseudo-overdominance is frequent, owing to multilocus epistatic interactions that are susceptible to rapid disassembly as a consequence of the unruly segregation and recombination that is expected to occur during selfing. In practice, therefore, the obvious course of action for corn breeders is to (1) isolate as many homozygous inbred lines as resources will permit (enough, it is hoped, to include at least some inbred × inbred pairs that will combine outstandingly well), (2) devise methods of identifying lines that combine well, while circumventing the dauntingly difficult task of crossing and testing every line with every other line, and (3) devise economically practicable methods of producing hybrid seed on a commercial scale.

Shull had perceived by 1909, on the basis of his genetic studies, that only very rare inbred × inbred combinations would be expected to yield more than their source populations even though the average inbred × inbred combinations would be expected to be no better than their source populations. This, ultimately, turned out to be precisely the case, and such hybrids should have the advantage of uniformity and reproducibility year after year. Nevertheless, Shull's proposal to produce corn commercially from inbred × inbred crosses was not put into use at the time. The main practical problem was that few inbreds of the time approached even one-half of the yield of open-pollinated varieties, so hybrid seed produced on the poor-yielding inbred seed parents would have been too costly. An additional problem was that seeds produced on inbreds were often misshapen and germinated poorly, thus further increasing seed costs to farmers.

Shull's proposal, made in 1909, to produce inbred × inbred hybrid seed did not, in fact, come into practical use until about 60 years later, in the early 1960s. Nevertheless, the first major use of hybrid corn in commercial production traces back to 1918, when D. F. Jones suggested making double-cross hybrids of the type $(A \times B) \times (C \times D)$. Double-cross seed, produced on vigorous high-yielding female single-cross plants, was economical and reproducible year after year, and double-cross hybrid populations were adequately uniform. This procedure proved enormously successful. Starting in the early 1930s, double-cross hybrids soon became predominant in U.S. corn production, and during the following decades double-crosses contributed to ever-increasing yields until they were replaced by still better $F_1$ single-cross hybrids in the early 1970s (Figure 5-1). Another major breakthrough of an economic nature in the production of hybrid corn had come in the 1960s. Until that time hybrid seed had been produced by mechanical detasseling of female seed parent plants that were interplanted with intended male parents. However, in the 1960s cytoplasmically sterilized female inbreds became available as a result of the efforts of many breeders, and they quickly replaced mechanical detasseling of the seed parents. This substantially reduced the cost of producing hybrid seed. An even more significant breakthrough came in the 1970s: the replacement of double-cross hybrids with single-cross hybrids. This substitution led to a succession of highly impressive improvements in adaptedness, yield, and other aspects of performance that have shown no sign of diminishing to the present (Figure 5-1).

It was recognized during the early stages of developing inbred lines that very few of the lines isolated from open-pollinated populations excelled in combining ability and, consequently, that it was likely that vastly large numbers of inbreds would be required to capture the full potential of the numerous source populations available. The problem was that selecting a single plant in each generation immediately imposed a new limit on progress. Evidently, what was needed was a breeding scheme or schemes that would maintain genetic variability in each population under selection and, at the same time, allow the frequency of desirable alleles and desirable combinations of alleles to be increased by providing for recombination within each source population. In the 1930s these ideas were developed into several related breeding systems that acquired the general title "recurrent selection." These systems, as used in corn, came to have the following operational features:

1. Plants from a heterozygous source were self-pollinated, and the progeny obtained were evaluated.
2. Plants with inferior performance were discarded, and the next generation was propagated from the superior plants.
3. The resulting intercross materials served, in turn, as source materials for additional cycles of selection and intercrossing.

Breeding schemes of this sort were apparently first suggested by Hayes and Garber (1919) and independently by East and Jones in 1920. However, data were not published and these early suggestions did not lead to use of the method. The first detailed description of this type of breeding scheme was published by Jenkins (1940), who reported the results of his experiments with early testing for what he called "general combining ability." However, it was not until 1945, when Hull suggested that selection after each of several cycles of intercrossing might be useful in improving combining ability, that the method acquired the name *recurrent selection*. According to Hull (1952), "Recurrent selection was meant to include reselection generation after generation, with intercrossing of selects to provide for genetic recombination." The advantage of this system was clearly that ultimate performance would be set not by the genotype of any single plant in any generation but by the most favorable combinations of alleles in large successive recurrently selected generations.

In discussing recurrent selection it is convenient to recognize four different types, based on the ways in which plants with desirable attributes were distinguished: (1) simple recurrent selection, (2) recurrent selection for general combining ability, (3) recurrent selection for specific combining ability, and (4) reciprocal recurrent selection. In simple recurrent selection, plants were divided into two groups, a group to be discarded and a group to be propagated further on the basis of phenotypic scores taken from individual plants or their selfed progeny. Because test crosses were not made, simple recurrent selection was essentially phenotypic mass selection with either a single or a 2-year cycle. This method was thus restricted to simply inherited characters with high heritability. The three remaining types of recurrent selection differed from simple recurrent selection in employing test crosses to measure combining ability. The value of any inbred rests ultimately on its ability to produce superior hybrids when crossed with other appropriate inbreds. In the early days of hybrid corn breeding, tests for combining ability (productivity in crosses) were conducted in a direct manner, that is, by crossing individual inbreds with as many other inbreds as practically feasible. Considering that $N(N-1)/2$ different single-crosses can be made from $N$ inbred lines (ignoring reciprocal crosses and selfs), it is apparent why this system of testing broke down as soon as substantial numbers of inbred lines became available for testing. With 10 inbreds, this direct method of testing required measuring the yielding ability of 45 $F_1$ hybrids; with 20 inbreds, 190 $F_1$ hybrids; and with 100 inbreds, 4,950 $F_1$ hybrids. The procedure finally adopted as standard practice in corn was suggested by Davis in 1927. Davis advocated the use of inbred × variety topcrosses to predict the general combining ability of inbreds. The correlation between inbred-variety topcross performance and average single-cross production—that is, between average combining ability and general combining ability—was ultimately found to be sufficiently high that it was considered safe to discard at least the lower-yielding half of the lines under test without serious risk of discarding valuable materials. The topcross proved to be a great boon in testing inbred lines because it permitted identification of the most promising inbred lines from a group of size $N$ lines instead of $N(N-1)/2$ $F_1$ crosses. Early testing of inbreds for general combining ability was first proposed by Jenkins. Originally, tests for combining ability had been deferred until the third, fourth, or even the fifth generation of selfing. Jenkins (1935) presented data suggesting that original selections ($S_0$) from source populations, or from first-generation selfed plants, differ in combining ability, and that differences could be detected by test crosses despite the problems posed by the heterozygosity of $S_0$ or $S_1$ plants. Subsequently, several other investigators presented data on the test cross performance of $S_0$ and later generation lines indicating that early testing indeed produces inbreds with good combining ability. However, other workers expressed doubts about the early testing procedure, holding that if discarding were done on the basis of topcross tests in the first or second inbred generations, many ultimately worthwhile inbred lines might be thrown away. Despite conflicting opinions, early testing came to be widely used to aid in eliminating plants or lines that seem to be unlikely to produce superior hybrids on further inbreeding.

After the more promising inbred lines had been identified on the basis of superior general combining ability, the next step was to predict those particular inbred × inbred crosses that were likely to produce the highest yields. This aspect of combining ability was designated *specific combining ability* by Sprague and Tatum in 1942. The inbred lines identified as having superior general combining ability (GCA) were then crossed in all possible pairs (diallel crosses) to create single-crosses, which were then evaluated in yield trials for specific combining ability (SCA). Thus, if all of the 10 possible single-crosses among five inbreds ($A$, $B$, $C$, $D$, $E$) were made, and cross $A \times B$ produced the highest yield, the $A \times B$ combination was considered to have the highest SCA. Whether two particular inbreds combine to produce a high-yielding single-cross depends not so much on the phenotypes of $A$ and $B$, but on the extent to which favorable alleles and favorable combinations of alleles from $A$ and $B$ complement each other—that is, on the phenotype of the $F_1$ hybrid. Experience also showed that inbreds derived from unrelated source populations from the same ecogeographical region often produced higher yielding inbreds than inbreds derived from related materials; thus, the relatedness of inbreds also turned out to be a significant component of heterosis in hybrids. Good inbreds are rare, and even the most outstanding inbreds nearly always yield < 50% of their $F_1$ crosses. The problem of predicting SCA obviously became much more complex in the period of double-crosses (mid-1930s to mid-1960s) than it had been earlier. The number of combinations in which four inbreds can be taken four at a time is $3N!$ $[4!][N-4]! = 14{,}535$ double-crosses, excluding reciprocals. Although ways were found to predict double-cross performance (largely using single-cross data), accurate evaluation of relative performance presented formidable problems, and many corn breeders postponed decisions to recommend particular combinations for production only after repeated field trials.

In summary, early hybrid corn-breeding programs usually featured four distinct stages: (1) phenotypic selection within and among variable populations, (2) topcross testing for GCA, (3) pairwise testing of inbreds for SCA, and (4) field testing of the more promising inbred combinations. Although corn yields had improved hardly at all from 1860 to 1930, the growing of double-cross hybrids starting in the 1930s led to great improvements in yield (Figure 5-1). In fact, 40 years of applying the technology described here in steps 1, 2, 3, and 4 resulted in inbred lines that had been improved to the point that single-cross hybrids became economically feasible by the early 1960s.

As hybrid-breeding technologies continued to improve, important changes took place in the ways breeders went about the task of screening for combining ability and fitting inbred lines into hybrid combinations. The single change that had, perhaps, greatest effect on the ways inbreds were developed featured redirecting inbred-line development away from developing entirely new inbred lines and toward making specific improvements in lines to be used as replacements in established hybrids. When an inbred line is being tailored to replace one of the inbreds in an established single-cross hybrid, the opposite inbred is the logical tester to use and testing procedures obviously should be designed expressly to correct the deficiencies of the inbred being replaced. Such procedures have advantages—for example, the GCA of potential replacement inbred lines can be evaluated by crossing with appropriate narrow-based testers, such as elite inbreds or related inbred lines, thus obviating laborious preliminary screening tests with genetically heterogeneous testers. Replacement inbreds not only can be developed more quickly, but are also easier to evaluate than completely new inbreds for characteristics such as superior roots, superior standing ability, superior stay-green capacity, and other desirable characteristics. The yield of corn in the United States doubled during the period of single-crosses (1960–1990). The proportion of this gain resulting from genetic improvement has been estimated in various studies to be more then 60%, whereas less than 40% was attributed to improved cultural practices, including increased use of fertilizer.

## DEVELOPING SUPERIOR SINGLE-CROSS HYBRIDS

As hybrid-breeding technology continued to advance, still more efficient ways of developing and breeding inbreds were found. A common observation was that inbreds isolated from unrelated populations from the same ecogeographical region often produced higher-yielding single-crosses than inbreds isolated from more closely related parental materials—that is, that the best hybrid within a given ecogeographical region was often obtained by crossing the two unrelated inbred lines with the best GCA values for that region. It thus became apparent that GCA is critically important in adaptedness and in yield. General combining ability results from the additive effects of many alleles of many different loci that cumulatively enhance performance in multiple microenvironments within a given ecogeographical region; thus, GCA better lends itself to improvement by repeated cycles of selection than nonadditive effects (SCA). Consequently, it appeared to be advantageous to determine GCA effects as precisely as possible for as many single-crosses derived from a single ecogeographical region as could be accommodated in reasonably sized testing programs. Topcrosses to an open-pollinated variety do, in fact, reduce the burdensome labor of making $N(N-1)/2$ single crosses; therefore, topcrosses to genetically heterogeneous open-pollinated parents were originally used to estimate GCA. However, single-crosses are more efficient than topcrosses in estimating general combining ability, and breeders soon recognized that estimates of the additive effects of GCA should be based on as many single-cross combinations as possible. The compromise that became popular was to estimate GCA from incomplete diallel crosses based on as many of the possible $N(N-1)/2$ $F_1$ combinations as could reasonably be accommodated. In practice, each inbred was crossed with perhaps half a dozen or so carefully selected inbred lines and tested in original evaluation tests. As testing progressed, poorer combinations were dropped and different lines substituted; when a number of superior combinations had been identified by testing, the testing was then likely to be extended to additional locations. Obtaining information on captious traits such as ear droppage, stock breakage, or barrenness often requires testing in unique environments. In present practice, determination of GCA may ultimately call for testing dozens of promising inbred lines.

Hundreds, or even thousands, of new inbred lines of corn are now developed and tested each year worldwide, but only a few are good enough to be used extensively. It has been estimated that only 10 to 20 elite inbred lines are widely used on a worldwide basis at any given time. The development of an elite inbred line is thus a very rare event. Continuing programs of hybrid improvement have, for many years, been advanced by substituting improved inbred lines as replacements for earlier predecessor lines. Although novel superior inbreds are now and then developed by fresh isolation of new inbreds from original sources, most corn breeders have become intensely aware that the chance of developing a new elite inbred from primitive materials is very small. Consequently, modern breeders usually start each new breeding effort with elite inbreds and follow a breeding strategy that preserves as much of the genetic makeup of the elite inbred as possible, while correcting whatever weaknesses prevented such elite inbreds from being still better. Accordingly, most modern breeders start with elite inbreds or with inbreds that are approaching elite status; thus, most new inbreds have come to be "fixed up" versions of proven or highly promising new elite inbreds.

The two most widely used breeding strategies for accomplishing the aforementioned desired purpose or purposes are the backcross and the pedigree methods. Many breeders backcross to the elite inbred parent line once, twice, or even more times to obtain segregating populations that resemble the elite line as closely as practicable before they start intense selection for traits to be modified. This allows breeders to focus their skills on retaining a satisfactory intensity of the trait under transfer, while recurrent backcrossing more or less automatically attends to the essential companion matter of recovering the numerous and often subtle features that elevated the elite inbred parent to its elite status. Natural selection in combination with breeder-directed selection in farmers' fields and breeders' nurseries practiced over large numbers of generations has clearly increased the frequencies of favorable alleles and of the more favorable lower-order combinations of alleles, at the same time decreasing the frequencies of less favorable alleles and less favorable lower-order combinations of alleles. But the presence of many favorable alleles and many favorable lower-order combinations of alleles is not enough. Higher-order combinations are no doubt very important too, and they usually develop slowly (see Chapter 10 and, especially, Chapter 11). Further, it is clear that once favorable multiallelic combinations have developed, it is important that such combinations not only be preserved but also that they be enhanced. The most effective way to preserve and enhance favorable multilocus combinations is to cross elite materials with elite relatives. Moreover, such crossing followed by extreme linebreeding is frequently preferred over sib crossing (especially as compared with wider crosses) because the wider the cross, the greater the chances that segregation will dismantle existing highly favorable multilocus combinations, especially if interlocus epistasis is important, as very frequently appears to be the case. The backcross strategy for improving inbred lines and their combining ability can be summed up as follows. This strategy strongly emphasizes keeping what is good, at the same time providing opportunities to introgress favorable exotic alleles into elite materials. The pedigree method is also widely used for these purposes. In corn the pedigree method usually features progeny tests of ear-to-row plantings in segregating generations usually derived from a cross between some elite inbred line and materials (often another inbred line) with compensating strengths, such as early flowering. The pedigree method frequently utilizes a single backcross made to the elite inbred parent; this is done because the first backcross generation (BCI) to the elite inbred parent provides outstanding opportunities to select for simply inherited traits. Typically, highly appealing BCI plants are selfed to produce $B_1F_2$ families in which selection is practiced, giving breeders optimum opportunities to exercise their skills in improving the general performance of lines ultimately developed by selfing. Crosses are often made among superior inbred lines, and further intercrosses are made among selects, often culminating in the crossing of superior surviving $F_5$ to $F_7$ lines and testing for superior combining ability of the resulting hybrid lines.

The intrinsic performances of inbred lines derived by both the backcross and pedigree methods have, in themselves, usually been superior to those of their earlier predecessor lines. More important, however, $F_1$ hybrids from such crosses have also been superior to the $F_1$ hybrids formed from earlier predecessor lines (Figure 5-1). Although additive effects associated with GCA are very important, specific combining ability is also important. Nonadditive gene action associated with SCA sorts out the best single-cross combinations for specific ecogeographical areas, combinations that can be formed from crosses among groups of closely related inbred lines from the same general ecogeographical area. Opportunities sometimes even arise in such segregating populations to select for favorable simply inherited characteristics that have been overlooked in earlier generations.

# FIFTEEN

# *Breeding Clonally Propagated Plants*

The crops that produce most of the carbohydrates and protein on which humans depend for nourishment are, for the most part, large-seeded self-pollinated grasses or legumes that complete their life cycles in a single year (Table 2-1). However, a second group of plants also contribute in major ways to human nutrition, namely, a number of clonally reproduced perennials. Although all of these clonal plants are at least potentially long-lived, the most important among them (Table 2-1), including white potatoes, cassava, yams, and sweet potatoes, are usually replanted each year from tubers or other stemlike plant parts and, hence, are managed as annuals in agriculture. At the opposite extreme on both size and durability scales are numerous long-lived rosaceous trees such as apples, plums, and almonds, as well as citrus, walnuts, and other tree crops that do not come into full bearing for a few to several years but thereafter may maintain high productivity for several decades. Many other important perennial crops are intermediate in persistence; for example, such reedy species as raspberries produce satisfactorily for only a few years before they must be replaced. Regardless of growth habit, all clones are potentially immortal. Some clones (e.g., clones of dates, grapes, olives) are known to have survived for long periods in cultivation (even for many centuries). However, clones of most crops, whatever their potential longevity, are unlikely to remain sufficiently productive to survive for more than a decade or two after establishment unless they can be kept free of infectious systemic diseases, which often are viruses. In recent years, with the development of improved cloning methods, especially sterile tissue-culture methods, more and more short-lived perennials are now managed agriculturally as annuals. Some true annuals are now also being propagated vegetatively, often with the use of tissue-culture methods. Such vegetative propagation provides not only planting materials that are identical genotypically but also the opportunity to start all individuals under carefully controlled and virtually the same cultural conditions. Such methods are increasingly popular in producing highly uniform products, especially vegetables and ornamentals. Most of the methods used in propagating vegetatively reproducible crops are ancient in origin, and these methods vary widely from crop to crop. Long-lived species, such as plums, almonds, citrus, and grapes, are nearly always propagated by budding or grafting a detached vegetative portion of a superior genotype (scion) on an appropriate rootstock.

Shorter-lived species are often propagated from lateral shoots (as with bananas), from tubers (as with potatoes), from leafy cuttings, or from $F_1$ or $F_2$ progeny of sexually produced $F_1$ hybrid plants (as with strawberries).

Nearly all clonally propagated species are perennial outcrossers in nature, and as a result of this mating system all such plants are highly heterozygous and conspicuously intolerant of inbreeding. This is presumably because only highly superior clones that carry complex favorably interacting complexes of alleles of many different loci, much like the genotypes of elite $F_1$ hybrids of corn, succeed in agriculture. Once the superiority of any clone has been identified, this superiority has typically been protected by continued vegetative reproduction, often over long periods of time. It apparently has also been widely recognized that when clonal materials reproduce sexually, whether by selfing or by outcrossing with other superior clones, the resulting progeny are likely to be much inferior to the parental clone or clones. The general approach to breeding superior clonal materials is therefore obvious. The breeder either selfs superior clones, or deliberately hybridizes superior clones, to generate populations of genetically variable individuals that thereafter are reproduced vegetatively once the better daughter clones have been identified and the poorer daughter clones have been eliminated. Vegetative propagation ensures that each vegetative descendent within each family will have identical genotypes from the outset. Thus, in one sense, clonal breeding is quick and easy because all of the genetic variability within each clone is fixed immediately as a result of vegetative reproduction. In practice, the breeding of clonal varieties usually is reduced to simply creating many heterozygous sexual progeny and selecting among different families to identify and isolate the "best" vegetatively reproducible clones. As selection among clonal families proceeds over additional vegetatively produced generations, only the better clones are allowed to survive and the numbers of individuals within the better clones are increased vegetatively.

Thus, it is possible that breeders will ultimately have in hand a substantial number of survivors of the very best clone or clones. The problems of selection therefore have certain similarities to those encountered by breeders whose goal is to develop homozygous monogenotypic inbred-line varieties of selfers or homozygous monogenotypic inbred lines of outcrossers that are to be hybridized to produce heterozygous but monogenotypic $F_1$ hybrid varieties. In producing pure-line varieties of inbreeders, or homozygous inbreds to be used in producing hybrid varieties, the problem of the breeder is to isolate the "best" homozygotes that develop as inbreeding progresses. However, the potential of a clonal family is fixed once it is formed during union of gametes from its heterozygous parents; hence, the task of the breeder is to identify what is already there and fixed in each clone derived from the pairwise union of gametes of its heterozygous parent or parents. The difference is that the potential of any family to be reproduced thereafter as a clone is fixed at the very outset (aside from rare mutations), whereas during the breeding of inbred line varieties or in developing homozygous inbred lines to be used in producing $F_1$ hybrid varieties, the potential of each family changes as inbreeding progresses and potentially superior nonsegregating combinations are assembled by selection during segregating generations.

The problems of selection in isolating superior clones nevertheless have features similar to those confronting the breeders of inbred-line varieties. In the first segregating generation, when selection is likely to be on a single-plant basis, selection will be confined largely to highly heritable characters. Thereafter selection gradually shifts to characteristics of lower heritability as sufficient numbers of plants of each clone become available so that increasingly effective interclonal selection can be practiced. During this period the numbers of clones will usually be reduced, as far as possible, by visual assessments before replicated trials are undertaken; should replicated trials be undertaken, they are usually repeated in different locations and/or seasons. Although early testing in replicated trials often helps to identify the better clones, it is often not done because it is expensive and time-consuming to carry out such tests sufficiently well to provide reliable estimates of the value of lowly heritable traits (such as yield), as compared with tests provided by "eyeball" appraisals by experienced breeders or farmers. This is particularly the case with long-lived perennials that have extended juvenile stages. Consequently, a frequent compromise is to grow a small sample of each of many different clonal families, sometimes in more than one location. Decisions are based mostly on visual appraisals of the small samples available to identify those clonal families to be discarded and those to be tested further (perhaps at other locations) or merely placed in a collection. This relatively quick and inexpensive method frequently gives experienced selectors a good idea of the potential of individual clonal families under a range of environmental conditions.

Similar arguments related to general combining ability (GCA) and specific combining ability (SCA) apply, not only to the choice of original parents but also to the choices of which among vegetatively produced families are to be saved once a breeding program is under way. There have been anecdotal reports that the original parents chosen because of their high yields, or above average performance for other difficult-to-measure traits, most often produce superior progeny and, hence, that breeders are more likely to be successful if they make heavy use of such parents and avoid poorer-performing parents. However, the goal is to produce, identify, and preserve the best combinations of alleles possible, and there have also been anecdotal reports of unpredictable positive SCA contributions to performance. Thus, breeders are often tempted to expand the number of parents and the number of families to be tested in the hope of finding serendipitous combinations. There are clearly hazards in restricting numbers of parents and/or the numbers of lines in later generations. But there are also hazards in using only very limited numbers of parents and evaluating only limited numbers of clonal isolates in later generations. Most of the evidence is anecdotal and often differs erratically from clonal species to clonal species, as well as from breeder to breeder. A formal experiment by Brown and Daniels (1973) answered the questions implied in the preceding discussion, at least in part, for sugar cane. A single large population of more than 600 clones, all raised from seedlings obtained by crossing superior clones, was established and mass-selected over three vegetative generations for two important characters, gross yield and sugar content. The mass-selected population responded to selection for both yield and sugar content, and after the three generations of mass selection, a number of subpopulations, each containing 30 to 50 promising clones, were made up. Each of these subpopulations exceeded the standard variety in sugar yield. Individual clones within the subpopulations were not isolated and evaluated; consequently, it was not clear whether the good yields measured in the clonal mixtures were attributable to a few outstanding clones in each mixture or whether the favorable yields of the mixtures were due to favorable phenotypic interactions among possibly less-than-ideal clones. Regardless, it became clear from this experiment that relatively inexpensive mass selection in clonal mixtures of sugar cane led to substantially improved yields overall in several clonal mixtures that were made up after three generations of mass selection.

## INBREEDING IN THE DEVELOPMENT OF CLONAL MATERIALS

In corn, sustained inbreeding (selfing) sometimes produces homozygous and monogenotypic lines which, when hybridized in pairs, produce at least occasional monotypic hybrids of remarkable uniformity and productivity. Although the genetic structure of any individual clonal plant is presumably much the same as that of any single plant of corn in an open-pollinated population, it seems unlikely that more than a few clonal species could stand inbreeding as intense as that breeders usually apply to corn (continued selfing) in developing homozygous inbred lines; there are anecdotal reports that most clonal species suffer rapid and severe deterioration on inbreeding. It has been suggested, however, that a moderate degree of inbreeding might be economically possible in some clonal species without incurring excessively severe losses through inviability or sterility.

Perhaps the most successful scheme with perennials, a scheme which is more or less parallel to the scheme used with corn, is that used with strawberries (*Rosaceae*). Strawberries are outbreeding octoploids; all plants are potentially highly variable genetically in each of the four sets of chromosomes that make up its four genomes. When different plants in cultivated varieties have been hybridized, extensive segregation inevitably occurs, which sometimes leads to superior individual plants. Strawberries are readily cloned through leafy cuttings, and such vegetative reproduction, of course, ensures that the characteristics of any desirable family will be preserved with great fidelity. It is also relatively easy and inexpensive to generate large clonal families, each such family genetically identical, from each different individual segregant. Because strawberries can be treated as annuals, it is relatively easy to determine the value of each family rather quickly by growing adequately large numbers of individuals of each clone under a range of different environmental circumstances. Thus, by generating very large numbers of clones and quickly assaying the worth of each clonal family, it has been possible to increase the performance of strawberries dramatically in recent years.

## REPRODUCTIVE DISORDERS

In outbreeding species that reproduce from seeds there is likely to be intense selection for reproductive normality in nature, and this is also the case in cultivation. Consequently, breeders rarely encounter either ovular or pollen sterility in annual or biennial species grown from seeds. However, in biennials such as sugar beets, and in many forage grasses or other species that are produced for a vegetative product, it is natural and usual for breeders to select against flowering because this discourages production of reproductive organs, thus maximizing the partition of biomass toward the desired vegetative product to be harvested. At the same time, however, adequate seed production must be maintained to meet the essential requirement of establishing the next generation economically; thus sterility or even partial sterility are not favored for cultivars that must be reproduced by means of seeds. This is also the case for perennials that are propagated clonally for their seeds (e.g., coffee, almonds) and for fruits that depend on seed set for development of the fleshy edible tissue surrounding the seed (such as peaches and apricots). It is thus not surprising that clonal perennials that are not grown for their seeds often suffer reproductive derangements. Yet at least some degree of reproductive derange-ment is present in almost all clonally reproduced crops. Cultivars of perennial species in which a vegetative product is paramount nearly always show some degree of reduced flowering, presumably developed as a negatively correlated response to selection for improved vegetative yield. Some cultivars of sweet potatoes and yams have never been induced to flower, and modern cultivars of white potatoes (*So-lanum tubersum*) usually flower much less than their less selected relatives. In many crops produced for vegetative parts, especially species grown for forage purposes, intraspecific and interspecific sterility barriers, including cytoplasmic sterility bar-riers, are superimposed on the aforementioned genetic tendencies to reduce seed production.

In summary: Normal reproduction often appears to be diminished in clonal crops grown primarily for some vegetative product. In such crops the breeder's task is rarely simple or straightforward. Regular diploid inheritance is often reduced, and formal genetic approaches are often difficult. The apparent simplicity of clonal breeding schemes, noted earlier, is thus often undermined by awkward and diffi-cult-to-handle practical realities. This is not to say that clonal breeding has not had numerous outstanding successes, but in many important species it is far from simple in practice. Perhaps the most positive aspect of clonal breeding is that of long-term constancy. Once clones have been reproduced vegetatively and carefully managed, particularly in respect to systemic viral diseases or unfortunate somatic mutations, they maintain their virtues over very long periods of time. Nevertheless, there is a positive as well as a negative aspect to mutation. Some mutations are desirable and can be fixed immediately and permanently; much of the diversity of old clonal crops has arisen as a result of spontaneous mutations with desirable effects that were noticed and preserved by farmers, nurserymen, and/or breeders. It has been suggested repeatedly that the usefulness of superior clones may be increased by directed alteration of genetic materials. Thus, desirable clones may be modified by molecular engineering to produce desired medicinal or industrial products less expensively than such products can be produced in factories. Of course, this may also be possible by genetic engineering of high-producing $F_1$ corn hybrids (or many other highly efficient agricultural plants) to produce industrial products. Still another possibility is to circumvent extraction of medicinals entirely by engineering clones—for example, clones of bananas that carry medicinals in concentrations such that proper dosages may be obtained by simply eating the fruit. These prospects have received publicity but have, apparently, had little impact so far.

# SIXTEEN

# Breeding Hybrid Varieties of Selfing Plants and Plants that Are Clonally Propagated in Nature

As we have seen in earlier chapters, the mating and reproductive systems employed by farmers and breeders after domestication have nearly always followed the system that prevailed in each species in nature prior to domestication. But breeding systems in plants are not immutable, and natural reproductive systems have occasionally been altered dramatically as breeders have discovered superior alternative breeding procedures. We now consider two such cases.

The earliest and perhaps the most striking modern change has been to replace the complex mixtures of highly heterogeneous multilocus genotypes that develop under open-pollination with single highly heterozygous but genetically and phenotypically homogeneous $F_1$ hybrid varieties. This was accomplished successfully in corn (Chapter 14) by developing very large numbers of highly inbred genetically homozygous lines, testing these lines to identify those few inbred lines that combine well to produce superior monogenotypic hybrids, and developing techniques to cross such lines in ways that permit production of $F_1$ hybrid seed inexpensively. Perhaps the most conspicuous trend at present is, in at least one sense, diametrically the opposite of the process by which hybrid corn was developed, namely, the replacement of homozygous pure-line varieties of inbreeding species with highly heterozygous but monogenotypic, and hence highly homogeneous and phenotypically uniform, $F_1$ hybrid varieties obtained by crossing homozygous inbred lines that "nick" to produce superior monogenotypic heterozygotes. Grain sorghums (*Sorghum vulgare*), usually ~ 95% selfing, provide an outstanding example of this shift in plant-breeding technology. However, a much earlier—indeed, very ancient—system in agriculture, has, in recent years, been revived with species such as strawberries that are relatively easily cloned. This system involves hybridizing clonally reproducible species and examining numerous sexually produced offspring for superior types, which can then be replicated asexually at will for testing to determine agricultural potential and commercial value. Grain sorghum provides another example of change in breeding methods, namely, converting near-homozygous selfing varieties to fully heterozygous $F_1$ hybrid varieties.

Grain sorghum yields doubled shortly after the release of the first $F_1$ hybrid varieties developed in the mid-1950s. At least half of the early hybrids were based on common varieties of the time. Webster (1976) suggested that, by 1960, the genetic potential of sorghums had nearly been exhausted because prehybrid cultivars traced back essentially to only 29 rather similar plant introductions. By the mid-1960s, however, a new generation of hybrids with much higher yielding ability had been developed, in large part stemming from introgressions of novel introductions of sorghums from the Sudan and Nigeria. These types supplied valuable traits such as drought resistance, tolerance to serious diseases, stiffer stalks, increased time from flowering to maximum dry weight (resulting in higher test weights), and, particularly, the development of dwarf types that were amenable to combine harvesting. By 1963 the world sorghum collection had been increased to more than 30,000 accessions. Perhaps more than two-thirds of this collection was photoperiod-sensitive for short-day length and most accessions carried dominant genes for tallness. The genetic variability needed for improvement, however, could not be utilized until the inception of the sorghum conversion to dwarf, long-day types program suggested by Stephens et al. (1967).

The process of conversion for sorghum began with crossing an early-maturing, day-length insensitive, genetically short (recessive for four height genes), temperate zone genetic stock with a tall, day-length sensitive sorghum introduction. These $F_1$ hybrids are then grown under short day-length, tropical conditions (e.g., in the Caribbean Islands) and self-pollinated during the winter, also under short-day conditions in the tropics. A large segregating $F_2$ population (1,200–2,000 plants) was grown under temperate zone conditions (> 35° north or south of the equator), and short-early plants were saved. The process then became one of recurrent backcrossing in a three-generation cycle such as the aforementioned, always using the plant introduction as the male. Thereafter, the plant introduction was used as the female parent to ensure that cytoplasm from the alien parent was present in the final step of the conversion. In the final step of the conversion several $F_3$ families descended from the fifth backcross generation were grown next to the original plant introduction parent under tropical conditions, and at maturity the family most similar to the original parent was selected for release. This process provided short, early (photoperiod-insensitive) converted lines that were ready for final screening by plant pathologists and entomologists for responses to disease and insect resistance and by breeders for suitability for use in crosses to produce commercial hybrid sorghum varieties. Although sorghum was the main recipient of the benefits of conversion, conversion programs and related procedures appear to open new opportunities for varietal improvement for other crops, as discussed by Maunder (1994).

We now consider another major change in breeding procedure. It has often been suggested that few, if any, clonally reproduced crops could conceivably be bred in any way other than as clones. The globe artichoke (*Cynara cardunculus*, Spp. *scolymus L.*) is a large thistle (*Compositae*) native to the Mediterranean Basin that has been grown for its large, fleshy heads for about 2,000 years. The globe artichoke has traditionally been propagated vegetatively from suckers, lateral shoots that develop from buds of the lower stem. Until very recently commercial production has been based on cultivation of such vegetatively propagated clones. According to

Basnizki and Zohary (1994), "Breeding activities in the last 30 years have focused on the following approaches: (1) collection of traditional, often local clonal cultivars, evaluation of their performance; (2) introduction of the more promising ones into new production areas; (3) evaluation of the aggregate of similar clones that frequently constitute a local cultivar and selecting among them the best performing clones; (4) breeding new cultivars by crossing different clones and selecting within promising hybrid progenies in attempts to identify selections better adapted to modern market requirements." However, their studies led these investigators to the conclusion that "traditional perennial, plantationlike cultivation, which requires heavy inputs of labor, is an inordinately costly operation." Because of the heavy investment necessary, growers usually maintain their plantations of artichokes for several years, even though yields and quality frequently decrease over time; furthermore, vegetatively propagated plantations are subject to damage from diseases and pests transmitted during vegetative reproduction. Unfortunately, the cost of labor in producing plants biotechnologically is also high, and field planting such plants is labor intensive. Consequently, Basnizski and Zohary (1987) began to experiment with seed-planted cultivation methods, and after several generations of self-pollination and selection, they developed a hybrid cultivar that could be seed-planted. Their breeding efforts then shifted to hybrid cultivars intended for seed planting. They concluded that the advantages of seed-planting globe artichokes also include (1) that mechanical sowing of seeds is a labor-saving, much less expensive operation than planting of suckers; (2) that with early fall planting, most artichoke genotypes bolt fully by the following spring (thus this perennial is readily convertible to an annual in agricultural practice); (3) that direct-seeded plants develop long, vertical taproots that capture moisture and nutrients more efficiently than planted suckers; and (4) that seed planting inhibits the transfer of soil-borne pathogens and pests and largely prevents viral infections, inasmuch as most plant viruses are not transmitted via seeds. These several elements all contribute usefully to more vigorous plant growth, healthier plants, and much lower costs for labor, fungicides, and pesticides.

In most cases the selfing of clonal artichoke cultivars has resulted in wide segregation in the first inbred generation. In some cases the effects of selfing were so severe that it was not possible to continue selfing beyond the third or fourth generation; however, in common with Pécaut (1993), Basnizski and Zohary (1994) concluded that selfed generations $S_3$ and $S_4$ provide practical and usable compromises between vigor, seed production, and homogeneity. To date, only genic male sterility has been found in the globe artichoke. Principe (1984) reported that male sterility governed by a single recessive gene ($ms_1$) had been detected in California, and Basnizski and Zohary (1994) have reported two additional nonallelic recessive male sterility genes, $ms_2$ and $ms_3$, which they have used in their breeding work. Pécaut and Foury (1992) examined 21 cross combinations between Inbred $I_1$ and Inbred $I_4$ parental lines; they found that the average yield of the $F_1$ hybrids was 81% higher than the yields in the parents. Basnizski and Zohary (1987) reported similar heterosis in crosses between cultivars and in $F_1$ hybrids extracted from such cultivars, and they also developed and released several seed-planted hybrid cultivars. The production technology developed for seed-planted artichokes has now been tested extensively in Italy and France (Basnizski and Zohary 1987); generally, the yields and quality of the seed-planted cultivars have been equal or superior to those of traditional cultivars (Basnizski and Zohary 1994).

These results indicate that a shift from traditional clonal production to seed planting is technically feasible in this previously clonally propagated plant, and it seems likely that such a shift will add seed-grown artichokes as a useful annual element in crop-rotation systems. It is also likely that in the near future breeders will focus on the development of a range of early- and late-seeded hybrid cultivars adapted to specific agricultural environments and that hybrids with concentrated yields, fit for mechanical harvesting, will soon be bred to meet the needs of the processing industry.

# SEVENTEEN

# *Plant Breeding for Low-Input Agricultures*

There is general agreement among plant breeders that interactions among geno-types and environments have an important bearing on methods of breeding and on the usefulness of the resulting cultivated varieties. Consequently, this discussion focuses on some vexatious aspects of genotype-environmental interactions that have arisen in recent years, particularly in areas of the world in which agriculture is practiced by resource-poor human populations who reside in generally harsh environments. Such human populations have, in general, not shared fully in the spectacular yield increases that have been made possible by modern breeding technologies, especially when these breeding technologies have been augmented with a variety of inputs, such as irrigation and improved mechanization, and/or with such inputs as fertilizers, pesticides, herbicides, and/or fungicides. In any case, problems of low yields in resource-poor areas remain a major challenge. It has been estimated (Pimbert 1994) that about 1.4 billion people worldwide depend on subsistence agricultures and that even though resource-poor farmers in such areas carry out about 60% of total global agriculture, they produce perhaps only 15 to 20% of the food of the world (Francis 1986). Breeding programs have, in general, been highly efficient in resource-rich areas, perhaps in large part because agricul-tural conditions in farmers' fields and in the agricultural research stations in which most new varieties are developed and evaluated are closely similar. However, genotype × environmental interactions frequently lead to serious problems in subsistence agricultures in resource-poor countries, where climatic conditions are often harsh and erratically unpredictable (Borlaug 1995; Ceccarelli 1994, 1997; Ceccarelli and Grando 1991). In subsistence agricultures yields in farmers' fields are often severalfold less than the yields realized at well-supported agricultural research stations. Low yields in farmers' fields in resource-poor countries often result from low levels of inputs, but also frequently from less-than-ideal agricultural practices. Yet it should be emphasized that less-than-ideal agricultural practices do not necessarily stem from ignorance; fertilizers, insecticides, and herbicides may not be readily available or may be too expensive for farmers in such areas. Moreover, the use of such inputs is often appropriately considered excessively risky by farmers in environments where the possibility of total crop failure is high, owing to unpredictable climatic crises such as those associated with drought or extreme temperatures. Further, the benefits of timely and/or recommended seedbed prepa-ration methods or other recommended agricultural practices may be irrelevant because needed equipment or supplies are too expensive and, hence, simply not available to farmers in such areas.

Resource-poor farmers in many areas of the world consequently often adopt strategies based on spreading the risk, such as through interspecific and/or intras-pecific diversity. Sometimes two or more crop species are grown in the same field at the same time (interspecific diversity), and/or two or more cultivars of the same species may be grown simultaneously in the same field (intraspecific diversity). Another level of intraspecific diversity is sometimes attained by growing genetically heterogeneous cultivars of the same crop. The latter stratagem is also sometimes used in advanced agricultures. In a few areas of California, particularly in small areas located downwind in the rain-shadows of mountains, areas that frequently suffer from localized severe moisture stress, many farmers have, of their own choice, saved seeds from originally heterozygous experimental populations of barley that had been grown experimentally in small plots on their farms (or on neighboring farms) for a number of years (see Chapter 5). The steadily improving performance (earlier flowering; earlier maturity; shorter, stiffer straw; and better grain yields) that developed over several years in such populations convinced many farmers (but not necessarily tradition-bound breeders) that the often self-selected mixtures of survivors in such populations were responsible for, and had in fact endowed such populations with, the ability to produce both higher and more dependable yields than had been provided by the traditional, more uniform, drought-resistant varie-ties they had grown previously. Although traditional plant-breeding methods usually appear to serve advanced agricultural systems well, it has become apparent that standard breeding methods are often not entirely in accord with the realities of low-input agricultures. The rest of this chapter will be directed to two aspects of this kind of situation in low-input agricultures: (1) identifying genotype-environ-mental interactions that have demonstrable bearing on the performance of differ-ing genotypes in disparate environments and (2) selecting in appropriately heterogeneous populations to develop populations (varieties) that are adapted in low-input environments. Thus, this chapter is concerned largely with ways of developing and taking advantage of favorable genotype × environment interactions ($G \times E$ interactions) in low-input agricultures.

## CLASSIFICATION OF GENOTYPE-ENVIRONMENTAL INTERACTIONS

As pointed out by Haldane in 1946, certain facts about genotype-environmental interactions are so simple they usually are not stated. Following Haldane, assume for illustrative purposes two genetically different populations, $A$ and $B$, and two differing environments, $X$ and $Y$. Further assume that measurements have been made on some important single character (say yield) and that statistically signifi-cant differences have been observed such that the three possible genotype × environmental combinations can be placed in rank order 1, 2, and 3, as shown in the graphical representations of Figure 17-1. The points to note about these

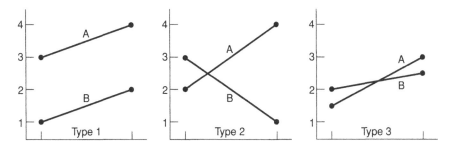

FIGURE 17-1  Graphical representation of three important types of genotype-environmental interactions

particular interaction types are (1) whether Genotype *A* does better than Genotype *B* in both environments (as in Type 1) and (2) whether Genotype *A* is superior to Genotype *B* in one environment but inferior in the second environment, as in Types 2 and 3. Note in particular that Types 2 and 3 are both crossover types; these kinds of interaction type are particularly important to plant breeders. In Type 1, Genotype *A* can be taken as the universal variety, whereas in Type 2, *A* versus *X* and *B* versus *Y* can be taken as superior and inferior specialized varieties and specialized environments, respectively, whereas in Type 4, also a crossover type, *A* and *X*, and *B* and *Y*, are less specialized varieties and less favorable environments, respectively.

   In actual practice the situation becomes immensely more complicated because breeders nearly always must deal with many genotypes and environments; if there are large numbers of intergrades in relative effects, the number of types of interaction can become very large (Allard and Bradshaw 1964). With only 2 genotypes and only 2 environments, and only a single criterion of classification, 4 different types of interaction are possible, but with 10 different genotypes and 10 different environments, 400 types of interaction are possible. Hence, interaction types rapidly increase to such large numbers that they become difficult to classify and their consequences and implications even more difficult to comprehend. Nevertheless, there is a main point that is worth noting. Improvements have been made in the adaptedness and the productivity of crop species in most, and possibly all, agricultural environments subsequent to their domestication. This indicates that the selection practiced by farmers and/or breeders has overcome at least some, and perhaps even many, of the myriad of difficulties that genotype-environmental complexities impose on progress, thus offering hope that further gains are likely in the future.

## TYPES OF GENOTYPIC-ENVIRONMENTAL INTERACTIONS

Variations in environments can be divided into two general categories: predictable and unpredictable (Allard and Bradshaw 1964). The first category includes those features of the environment that fluctuate in very regular fashion at a given location, such as day length (photoperiod) and some that fluctuate less regularly, such as day-to-day maximum and minimum temperatures. Also included are those aspects of cultivated environments that are largely determined by farmers, such as planting time, sowing density, and other agricultural practices, which often involve interactions that can be adjusted more or less at will. The second category includes short-term fluctuations in weather, such as the amounts and distribution of rainfall and damaging temperatures (hot or cold) that occur more sporadically and are often beyond control or avoidance. In poorly mechanized agricultures this second category may also include variations in agricultural practices that may possibly sometimes be held appropriately constant in better-mechanized situations. The distinction between the predictable and unpredictable categories is far from clear-cut. Nevertheless, the various qualifications that are appropriate should not be allowed to conceal the differences between the two categories. Whenever there are significant variety × location interactions between different genotypes or varieties in a particular geographical region, this indicates that potentially important environmental differences exist in the region. However, great precision in the conduct of any one trial at any single location has often been found to be wasteful of effort. For example, in an assessment of spring oat varieties carried out in Great Britain, which has relatively predictable within-season weather conditions, it was found that one replication was usually adequate and that two replications were fully sufficient in trials conducted in any one place. During the 1950s double-cross corn hybrids, which are moderately well buffered against environmental fluctuations, were usually tested at only 5 or 10 locations; later the number of locations was increased to 15 to 20, at the expense of replications (Sprague and Federer 1951). At present, advanced corn-belt single-crosses are usually tested at more than 100 single-replication locations; more locations identify widely adapted single-crosses, as well as more "adaptedness" genes (Troyer 1997). Crossover-type interactions are rare for single-cross maize hybrids because stability regressions usually range from only about 0.95 to 1.05 for elite single-cross corn hybrids; moreover, farmers in the corn belt expect yields to be consistently high, higher than 2.0 tons/ha (Troyer 1997). Modern single-cross varieties of corn are highly favored by corn-belt farmers because such varieties consistently produce exceptionally well in nearly all years; they are also favored by seed companies because the sale of large amounts of seed of any single variety is economically beneficial to both farmers (low seed costs and high yields) and seed companies (large seed sales of relatively few superior hybrids). There is a worldwide trend in corn toward fewer, more widely adapted, single-cross hybrid varieties. These corn varieties have resulted from very wide area testing and from utilizing high-plant-density stresses in screening hybrids (Troyer 1997).

Agriculture in many resource-poor countries, particularly under unfavorable conditions, presents a very different situation than that described in the preceding paragraphs. Yield levels in farmers' fields are often several times lower than those obtained at breeding stations, and yields often vary widely from year to year. Breeding carried out at high-input research stations may fit advanced agricultural situations very well, but such breeding often turns out to be less compatible with the realities of low-input agricultures as practiced by resource-poor farmers. The rest of this chapter will be concerned with the two main factors on which progress depends in both sets of circumstances: (1) the type of germplasm on which selection is practiced and (2) the environment(s) in which the experimental materials are grown, selected, and/or evaluated. Thus the theoretical basis for progress lies in the genetic potential of the original germplasm chosen to initiate a breeding program and the genotype-environmental interactions to which chosen materials are exposed during both selection and evaluation.

## INBREEDING

From this point on we focus attention on inbreeding plants, because the most important cultivated plants are inbreeders (Table 2-1) and because inbreeders offer many advantages in genetic studies. The term *inbreeding* is applied when the individuals that mate are more closely related than random members of indefinitely large populations. In natural as well as cultivated populations there are two general causes of inbreeding: (1) restriction in population size and (2) mating-system differences among the individuals in the population. In a population of bisexual organisms each individual often has two parents (nearly always only one parent in the case of selfers), four grandparents, and so on, so that in $N$ generations back any individual in an outcrossing population may have many different ancestors, but in a heavily selfing population of annual plants, any individual may possibly have only one ancestor. The number of individuals required to provide separate ancestors for each individual in a given generation obviously becomes very large within a few generations, so all individuals within any population (including a selfing population) are likely to have many common ancestors in the not-so-remote past. The smaller the population, the less remote the common ancestors are likely to be and, consequently, the higher the level of inbreeding. Inbreeding resulting from restricted population size is not confined to populations consisting of only a few individuals isolated from others of their own kind. Many factors can lead to effective restriction of population size, even though the total number of individuals in a population may be very large. Isolation by distance provides an important example: neighbors are likely to mate, and proximity in space and/or time also increases the probability that close neighbors will be relatives. The other main sources of inbreeding derive from various mechanisms that affect the mating system. In plants, modifications of floral structures that encourage self-fertilization are very common, and in many instances such mechanisms are remarkably effective, reducing the amount of outcrossing to 1% or, occasionally, even less (Chapter 4). It is often argued that inbreeding, by increasing homozygosity, forces populations of inbreeders to achieve very close adaptedness to their immediate environments. It is also frequently argued that gain in adaptedness (fitness in the local environment) is achieved largely through loss in adaptability (flexibility or capacity to change in adaptedness) as homozygosity increases. Such speculations have also been the basis for numerous conjectures concerning evolutionary changes. There is currently a large body of evidence indicating that selection is particularly effective in improving performance in self-pollinated plants (Roberts 1929). Harlan (1957), in discussing the variability he observed in barley (usually ~99% selfed) across the Eurasian landmass, stated that within each local area there has evolved a type particularly fitted for the conditions that exist there. Slight changes in altitude are accompanied by corresponding changes in the barleys. The barleys from each tiny area are made up of large numbers of strains that look much alike but that are made up of many strains that differ in ways useful to the plant breeder. Even out on the plains where the superficial appearance of the crop may be the same over a large area, these constituents are present in endless variety and they shift as one goes from drier or colder sections.

Thus, it came to be recognized that populations of cultivated plants and their wild progenitors often differ in important ways in adaptedness to the environment. It is only recently, however, that detailed studies of quantitative variability have been undertaken and that adequate estimates have been made of the components of such variability under various environmental conditions. In these studies the basic procedure has usually been to collect seeds from many different single randomly chosen plants of a single species from ecologically differing areas. The seeds are then sown in replicated experiments in one or more environmental situation, and individuals within each family are measured for various morphological and physiological attributes. Data from these studies have often permitted quantification of three aspects of genetic variability: (1) those associated with broad geographical regions, (2) variability within specific sites within regions, and (3) variability associated with very small local areas. Recall from Chapters 10 and 11 that such differences can often be characterized very precisely in natural populations of *Avena hirtula* and in *A. barbata* by scoring the frequencies of simply inherited, electrophoretically detectable molecular markers that presumably identify short chromosome segments, each of which carries at least one allele (and often more than one allele) of each of the many loci affecting adaptedness. The data of Chapters 10 and 11 indicate that the population genotype of selfers thus appears to result from complex interactions, often epistatic, among many genotypes, giving each population a coordinated and cohesive structure. Each population appears to prosper in the specific habitat it occupies, owing to harmonious balances that build up among the several genotypes of which the population is composed; each population thus appears to have favorable integrative and cohesive properties resulting from favorable interactions and interdependencies among its constituent genotypes. The experiments of Akemine and Sakai (1958) and Allard and Jain (1962) revealed that stabilizing selection and directional selection were both important in heavily selfing cultivated rice, as well as in heavily selfing cultivated barley populations. Akemine and Sakai grew hybrid rice populations at 20 experiment stations located at 31° to 43° north latitude in Japan and after many generations drew random samples of seed from the populations that had been grown at each location. The seeds were then grown in a common environment in central Japan, and each population was measured for various characters. The data revealed particularly strong directional selection for earliness and for lateness at northern and southern sites, respectively, whereas there was much less change in mean heading time for populations grown in intermediate latitudes. However, at least some directional selection, as well as much stabilizing selection for yield, took place at all locations. Similar concurrent directional and stabilizing selection responses have also been observed in experimental populations of highly selfing populations of barley, wheat, and *Phaseolus* beans in California. For example, Allard and Jain (1962) observed that the mean heading date in Composite Cross V of barley, synthesized by mixing the $N(N-1)/2 = 465$ possible $F_1$ intercrosses among 31 barley varieties (most of which were later in maturity than optimal for the test environment), shifted rapidly in the direction of earliness in generations $F_2$ to $F_4$ and continued to shift but more slowly, in the next 15 generations. In generations $F_2$ to $F_4$ many individuals were conspicuously inferior to locally adapted varieties. However, even in the absence of artificial selection, seed yields improved rapidly thereafter (18% increase from generations $F_4$ to $F_{14}$ and an additional 5% increase from generations $F_{14}$ to $F_{21}$. In generation $F_{24}$ the mixed population was found to yield significantly more than the pure-line local variety recommended for the experimental area. These results indicate that the fitness of populations derived from crosses among many arbitrary parents is often low in early generations owing to the production by segregation and recombination of numerous unbalanced genotypes. But the evolution of such populations is frequently characterized in early stages by prompt elimination of inferior types, leading to rapid improvement in mean population fitness. Continued evolution of each system often appears to feature incorporation of genotypes with slightly differing genetic characteristics, genotypes that appear to cooperate in giving the population a coordinated and cohesive structure; such changes are also often accompanied by increasing phenotypic uniformity of the constituent genotypes, particularly for characteristics such as height, flowering time, and time to maturity.

The *Festuca microstachys* complex of wild grasses provides particularly favorable opportunities for studying the long-term evolution of local population structure under extreme inbreeding, because the strict cleiostogamous habitat of these annual wild grasses virtually ensures nearly complete self-fertilization, >99% (Kannenberg and Allard 1967). Quantitative studies of genetic variability within and among populations occupying distinctly different habitats revealed that these grasses exhibit gradually shifting adaptive patterns of variation in thousands of differing habitats, including ecologically intermediate areas located between distinctively differing areas, areas that these grasses have occupied over long periods of time across most of western North America from Mexico into Canada. Variation was frequently observed in close and obvious associations with rainfall, temperature, and edaphic features of the physical environment. Moreover, distinctive, broad geographical variations were superimposed on a fine patchwork and/or short-distance microgradient patterns that clearly mirrored relatively minor, but nevertheless unambiguous, progressive changes in the local environment. Many such short-distance gradients were also observed on gradual slopes, from the tops down the sides of hillocks, in which slope and exposure to the sun were clearly responsible for short-distance but readily perceivable differences in phenotype clearly associated with degree of aridity. These fescues were shorter and earlier in maturity in the steeper places on southern exposures that were most directly exposed to the sun; however, populations located lower on slopes that were less directly exposed to the sun and that received runoff water from above were noticeably more lush. Progeny tests conducted under uniform experimental conditions established that these differences were heritable and that, as expected from the mating system, most plants were fully homozygous. Plants from closely adjacent locations in nature often differed slightly but significantly in quantitative characters such as height. Later, with the advent of electrophoretically detectable variants, it also became clear that individuals that were phenotypically nearly identical often had distinctly different genotypes. The maintenance of numerous genetically distinguishable homozygotes at the molecular level in populations of these intensely inbreeding grasses of the *Festuca microstachys* complex appears to depend, at least partly, on interactions among individuals with unambiguously distinguishable genotypes, as well as on interactions among individuals that differ, slightly to even substantially, for quantitative characters such as height. It became clear from such studies that plants with intermediate measurements nearly always evolved in such intermediate habitats. Plant breeders and quantitative geneticists have usually regarded adaptation as the process of becoming more adapted and adaptedness as the ability to perform well in a given environment. Breeders have nearly always relied on analysis of variance of yield trials to identify the varieties that do best in the environments in which specific crops give economic yields. But, as Charles Darwin (1868) noted, "Plants are more strictly adapted to climate than animals. With every plant long cultivated, varieties exist for different climates. Adaptation to climate is often close." Yet most crop plants can produce acceptable economic yields in many different environments, which introduces a major problem for plant breeders: it is unrealistic to attempt to breed a single "best" variety for each environment. Borlaug (1957) provided one solution to this problem somewhat accidentally when he started the "shuttle" breeding approach at CIMMYT with the objective of speeding up breeding for stem rust resistance. One of the important results of moving germplasm rapidly from place to place was that it led to selection for photoperiod-insensitive wheat genotypes; this permitted wide spread of CIMMYT wheat cultivars throughout much of the world. That is, CIMMYT's wide-testing strategy led not only to the development of many widely grown wheat varieties, but to the definition of *megaenvironments*, and to the use of very wide testing in many other crop species. A megaenvironment is defined as a "broad not necessarily continuous often transcontinental area with similar biotic and abiotic stresses, cropping systems, and consumer preferences (Braun, Pfeiffer, and Pollmer 1992). Troyer (1997) posed the question of how best to identify superior widely adapted varieties; his answer, for corn, was, "Test widely." Gauch and Zobel (1996), in their comprehensive analysis entitled "Identifying Megaenvironments and Targeting Genotypes," stipulated that it is useful to subdivide a region into several relatively homogeneous megaenvironments and to breed and target adapted genotypes for each megaenvironment. They stated as goals the identification of relevant criteria for evaluating megaenvironments and applying the Additive Main Effects and Multiplicative Interaction (AMMI) statistical model to megaenvironment analysis. According to these authors, statistical strategies for identifying megaenvironments should meet four criteria: (1) flexibility in handling yield trials with differing designs, (2) a focus on that fraction of the total variation that is relevant to identifying megaenvironments, (3) duality in providing integrated information on genotypes as well as environments, and (4) relevance to showing which genotypes are best in which environments. They stated that preliminary results with the AMMI model indicate that a small and practicable number of megaenvironments often suffice to exploit useful interactions and increase yields. Braun et al. (1992) reported that low-productivity environments are prone to large errors, less differentiation between genotypes, and less repeatability across years. Nevertheless, Ceccarelli and Grando (1993) found that selection of locally adapted barley landraces in low-yielding megaenvironments in at least one case identified superior entries that outyielded the check variety 30 times more frequently than was the case with selection of elite barley lines in a high-yielding megaenvironment. In addition, Gauch (1988) and Gauch and Zobel (1996) reported that AMMI analyses of regional data "can help in controlling errors and in gaining accuracy." Such genotype-based, as well as environment-based, proposals are interdependent and complementary (Peterson and Pfeiffer 1989). The genotype-based concept appears to be relevant because planting the winning genotype in each megaenvironment maximizes yield; the environment-based idea is relevant because planting the genotype that wins in each environment is helpful in assigning each local area to the appropriate megaenvironment and, hence, to the appropriate genotypic recommendation. Gauch and Zobel's title, "Identifying Megaenvironments and Targeting Genotypes," implies that the two concepts are intrinsically interwoven and interrelated. This view is consistent with the view of Darwin (1859), namely, that "selection is the chief agent of evolutionary change," which was mentioned in the first sentence of this book. It seems appropriate to bring the second edition of *Principles of Plant Breeding* to an end with another quotation from the *Origin of the Species*: "With every plant long cultivated varieties exist for different climates; adaptation to climates is very close" (Darwin 1859).

# *Glossary*

**Adaptedness**—The degree of suitability of an organism for its environment, developed over time as a result of selection.

**Allozymes**—Allelic forms of enzyme loci that can be distinguished by electrophoresis.

**Allele**—Any one among many possible variant forms of genes that result from mutations.

**A posteriori**—Derived by reasoning from observed facts (data).

**A priori**—Based on hypothesis or theory rather than on experiments or experience.

**Asexual reproduction**—Reproduction by processes that do not involve union of gametes.

**Biometrical genetics**—A branch of genetics concerned with the application of statistics to biological problems, especially the study of quantitative genetics.

**Biometry**—Statistical analysis of biological observations.

**Blending theory of inheritance**—Early disproven theory that the genetic elements of the parents fuse during fertilizations and lose their purity.

**Breeding (plant and animal)**—The controlled evolution of plants and animals by humans with the goal of producing populations that have superior agricultural and economic characteristics.

**Chasmogamy**—Maturation of flowers before the anthers burst, promoting cross-fertilization.

Note: The rapidly expanding vocabulary of genetics and plant breeding exerts great pressure on the author to choose words to be included in a glossary. This glossary is, accordingly, restricted to words that facilitate reading the text without reference to an unabridged dictionary of the English language. Nearly all technical terms used in the text of *Principles of Plant Breeding* that do not appear in this glossary can be found in abridged dictionaries such as the *American Heritage Dictionary of the English Language* (Houghton Mifflin Company, 1978), the *Merriam-Webster Collegiate Dictionary* (Tenth edition, 1995) or the *Third Edition of a Dictionary of Genetics* (King and Stansfield, Oxford University Press, 1985).

Clade—A group of related biological taxa that includes all descendants of an often remote common ancestor.

Cladistics—A system of biological taxonomy that reconstructs phylogenies in terms of successive sequences of branching ancestor-descendant lineages.

Cladogenesis—Evolutionary change characterized by treelike (dendritic) branching, illustrating phylogenetic relationships.

Cleistogamy—Flowers that remain closed until after anthers burst so that fertilization by selfing is likely.

Coadaptation—The selection processes that tend to accumulate harmoniously interacting (coadapted) individuals or (coadapted) alleles into the gene pool of populations; also the selection processes that tend to favor individuals that have mutually beneficial phenotypic associations with each other in populations.

Correlation coefficient—A measure of the degree of association between two or more variables.

Cross-pollination—Pollination by transfer of pollen from one plant to the stigmas of another plant; also called outcrossing.

Cultivar—A distinctive agriculturally derived cultivated variety of plant produced by selective breeding by humans and maintained by humans.

Deoxyribonucleic acid (DNA)—A chromosomal constituent of all living cell nuclei, consisting of two long chains of alternating phosphate and ribose units twisted into a double helix and joined by hydrogen bonds between the complementary bases adenine and thiamine, or cytocine and guanine, bonded in a sequence that determines individual hereditary characteristics.

Dioecious—Plants in which staminate and pistillate flowers are borne on different individuals.

Electrophoresis—Directional movement of charged particles in a gel placed in an electrical field.

Endonuclease—An enzyme that cleaves within a polynucleotide chain.

Environment—The complex of physical and biotic factors within which an individual organism or a population exists.

Epistasis—The property of nonreciprocal modification of the manifestation of one gene, said to be hypostatic, by the action of another gene, said to be epistatic.

Evolution—Descent with modification resulting from selection. The primary elements of biological evolution are (1) mutations of individual genes that supply the raw materials, (2) segregation and recombination following sexual reproduction to produce novel multilocus genotypes, and (3) selection preserving those genotypes that are better fit in the habitats they come to occupy.

Exon—Sequence of base pairs in a gene that participates in the coding of peptides.

Expressivity—*See* penetrance.

Fertilization—Fusion of the nuclei of male and female gametes.

Filial generation—Any generation following the parental generation. The generation resulting from a cross of parental individuals is called the $F_1$, or first filial generation; the selfing or intercrossing of $F_1$ individuals leads to the $F_2$, or second filial generation. The progeny of $F_2$ individuals make the $F_3$, or third filial generation (method of propagation normally specified).

**Fitness**—The relative ability of alleles, multigenic combinations of alleles, individuals, or populations to survive and transmit their genotype to following generations.

**Fossil**—A remnant or an impression of an organism preserved from a past geologic age.

**Genealogy**—A record of the descent of an individual, or group of individuals, from an ancestor or ancestors (or from older forms).

**Genetic drift**—Random fluctuations in allelic or genotypic frequencies resulting from small population size.

**Genotype**—The genetic constitution of an organism as distinguished from its appearance (phenotype).

**Geologic time**—Divisions and history of life on earth.

**Glossary**—A partial dictionary of a particular subject, explaining words, terms, and concepts.

**Hardy-Weinberg rule**—A statement in mathematical terms concerning the relationships between allelic and genotypic frequencies for random-mating populations in the absence of mutation, migration, and selection.

**Heredity**—The processes that tend to make progeny resemble parents and other ancestors.

**Heritability**—The proportion of the observed variability that is due to heredity, the rest being due to environmental causes. More strictly, the proportion of the observed variability owing solely to all additive effects of genes.

**Heterosis**—Vigor, such that genetically diverse individuals fall outside the range of their parents in respect to growth or productivity; historically attributed to heterozygosity but more recently shown to be due primarily to epistasis.

**Heterozygous**—Individuals in which the two alleles at corresponding loci differ for one or more loci.

**Homozygous**—Individuals in which the two alleles at corresponding loci on homologous chromosomes are identical for one or more loci.

**Inbreeding coefficient**—A measure of the probability that any two alleles of a locus will be identical by virtue of descent from a common ancestor.

**Incompatibility**—Failure of self- or cross-fertilization owing to antigenic differences that act between pollination and fertilization.

**Intron**—A sequence of a gene that is transcribed into nuclear RNA but is removed from within the transcript and rapidly degraded.

**Isogenic**—Lines that have been made identical at all loci by continued backcrossing (or selfing) while holding one (or very few) loci heterozygous and selecting alternative alleles at such loci.

**Isozymes**—Multiple forms of a single enzyme.

**Iteration**—A procedure that makes use of repeated trials to find the best-fitting value of a parameter from observed data.

**Landrace**—Any distinctive race of a cultivated species that has become genetically differentiated as a result of natural and/or human selection operating in ecologically different circumstances in the various areas to which cultivated species have been transported and have gained a foothold.

**Linebreeding**—The mating, usually in successive generations, of individuals having a known common ancestor.

**Linkage**—Association in inheritance of genes and the characters they control owing to location of genes of different loci in proximity on the same chromosome.

**Linkage map**—Map of the position of genes in chromosomes determined by recombination relationships.

**Linkage value**—Recombination fraction expressing the proportion of crossovers versus parental combinations in a progeny.

**Malthusian**—Relating to Thomas Robert Malthus and his theory that population numbers tend to increase at a faster rate than means of subsistence, leading to competition for environmental resources in short supply.

**Mating**—In eukaryotes, the pairwise union of unisexual individuals for the purpose of producing zygotes. There are four general types of departures from random mating: (1) genetic assortative mating, (2) phenotypic assortative mating, (3) genetic disassortative mating, and (4) phenotypic disassortative mating.

**Matriclinous**—Having predominantly maternal hereditary traits.

**Maximum likelihood**—A statistical procedure for estimating values of population parameters from sample data; the method identifies values that have maximum probability of being the best fitting for any given set of observations.

**Meiosis**—A double mitosis preceding sexual reproduction that results in the production of gametes with haploid (*n*) chromosome number.

**Mesolithic Age**—The cultural period between the Paleolithic and Neolithic Ages; appearance of the bow and cutting tools.

**Mitosis**—The process by which a nucleus is divided into two daughter nuclei with equivalent chromosome complements, usually accompanied by division of the cell containing the nucleus.

**Modifying gene**—Any gene that affects the phenotypic expression of genes at other loci.

**Molecular biology**—A branch of modern biology in which biological phenomena are studied by physical, chemical, and biochemical investigations at the molecular level. Molecular genetics features the study of those aspects of genetic systems that can be described at the molecular level.

**Mutant**—Any detectable and heritable change in genetic material not caused by genetic segregation and/or recombination.

**Mutation**—The processes that lead to heritable structural changes in genes.

**Naturalist**—A student of field biology, in contrast to a laboratory worker.

**Neo-Darwinian evolution**—A modified Darwinian paradigm developed during the 1930s and 1940s that incorporated concepts from modern population, ecological, and evolutionary genetics into descent with modification; often referred to as the Modern Evolutionary Synthesis.

**Neolithic Age**—Cultural period beginning ~12,000 years ago; appearance of technically advanced stone tools.

**Paleolithic Age**—Cultural period beginning with the earliest chipped stone tools (~750,000 years ago) until the beginning of the Mesolithic age, ca. 15,000 years ago.

**Parthenogenesis**—Reproduction by development of an unfertilized gamete, usually a female gamete, that occurs especially among some lower plants and animals.

**Penetrance**—The frequency with which an allele of a single locus or alleles of several loci are manifested on the phenotype of the carriers. Penetrance, as well as expressivity, often depends on both the genotype and the environment.

**Peptide**—A compound formed from two or more amino acids.

**Phenotype**—The observable properties of an organism brought about by its genotype in concert with the environment in which the organism develops.

**Phyletic evolution**—Gradual transformation of one species into another without branching.

**Pleiotropy**—The phenomenon whereby a single gene appears to be responsible for more than a single seemingly unrelated phenotypic effect.

**Prokaryote**—The kingdom (including viruses, bacteria, and blue-green algae) lacking membrane-bounded eukaryotic organization of the genetic material.

**Qualitative character**—A character in which genetic variation is discontinuous.

**Quantitative character**—A character in which variations are continuous so that classification into discrete categories is not feasible.

**Quantitative genetics**—The genetics of quantitative characters (continuously varying characters).

**Random mating**—A population-mating system in which each male gamete has an equal opportunity to fertilize each female gamete.

**Recombinant inbred lines (RIL)**—A set of inbred lines, each derived independently, usually by selfing single plants within lines during a number of filial generations. In plant breeding individual RIL are often derived by single-seed descent from a backcross of the $F_2$ to one or the other parent. Each RIL is ultimately expected to have a different pattern of alternative alleles of multiple loci within an otherwise unique homozygous genetic background, allowing effects of individual alleles (or combinations of alleles) to be evaluated in an otherwise homozygous background.

**Recombination**—Formation of new combinations of genes as a result of segregation in crosses between genetically different parents.

**Restriction fragment**—A fragment of DNA created by cleavage at specific sites by a restriction endonuclease.

**Restriction fragment length polymorphism (RFLP)**—Variations occurring within a species in the length of DNA fragments generated by a specific endonuclease. RFLPs are useful as molecular markers.

**Ribonucleic acid (RNA)**—A constituent of all living cells, consisting of a single-stranded chain of alternating phophotase and ribose units with bases adenine, guanine, cytocine, and uracil bonded to the ribose; the structure and base sequence of RNA determine protein syntheses.

**Segregation**—Separation of maternal from paternal chromosomes at meiosis and consequent separation of genes, leading to the possibility of recombination in the offspring.

**Selection (biological)**—Any nonrandom process that causes individuals with different genotypes to be represented unequally in subsequent generations. Artificial selection is a purposeful process with goals set by breeders.

**Self-pollination**—Transfer of pollen from an anther of a flower to the stigma of the same flower, or from another flower of the same plant or clone; contrasted with cross-pollinating or outcrossing plants.

**Species (biological)**—An aggregation of populations that are capable of interbreeding more or less freely but behaving as a distinct phyletic lineage evolving separately from other lineages. Species are often subdivided into subspecies, ecotypes, varieties, or other subspecific categories.

**Sport**—In plant or animal breeding, an aberrant individual resulting from mutation.

**Statistics**—A branch of mathematics dealing with the collection, analysis, and interpretation of numerical data.

**Stoma** (*plural* **stomata**)—A minute opening in the epidermis of a plant organ (as a leaf) through which gaseous exchanges occur.

**Supergene**—A group of genes held together mechanically by various mechanisms and inherited as a single unit. If coadapted, such genes, even though not necessarily functionally related, may cooperate to produce some adaptive characteristic.

**Taxon** (*plural* **taxa**)—A group of similar organisms that share a set of characters that are considered sufficiently distinctive to be worthy of a formal name.

**Test cross**—The cross of an individual of unknown genotype to an individual known to carry only recessive alleles of loci in question for the purpose of determining the genotype of the unknown parent.

**Textbook**—A book giving a systematic presentation of the principles and vocabulary of a particular subject.

**Threshold effect**—A term usually applied to traits with a polygenic basis that develop when the dosage of contributory alleles exceed a critical value in particular environments (sometimes used to explain all-or-none phenomena based on polygenically inherited characters such as resistance versus susceptibility to some diseases).

**Topcross**—A cross between a selection, a line, or a clone and a common pollen parent, which may be a variety, a single cross, or a number of elite inbred lines. The common pollen parents are called the topcross or tester parents. In corn, topcrosses are commonly inbred-variety crosses.

**Transposon**—A segment of genetic material that is capable of changing its location in the genome, especially when it contains genetic material controlling functions other than those needed for its insertion.

**Uniparental**—Having only a single parent—parthenogenetic.

# References

Akemine, H., and R. Sakai. 1958 (in Japanese). pp. 89–105. In: R. Sakai, R. Takahashi, and H. Akemine (eds.), Studies on the bulk method of plant breeding.

Allard, R. W. 1965. pp. 49–76. In: The genetics of colonizing species. Academic Press, New York.

Allard, R. W. 1967. Ciencia e cultura 19: 145–150.

Allard, R. W. 1975. Genetics 79: 115–126.

Allard, R. W. 1988. J. Hered. 79: 225–238.

Allard, R. W. 1990. J. Hered. 81: 1–6.

Allard, R. W. 1996. Genetic basis of the evolution of adaptedness in plants. pp. 1–6. In: P. M. A. Tigerstedt (ed.), Adaptation in plant breeding. Kluwer Academic Publishers, The Netherlands.

Allard, R. W., and Julian Adams. 1969a. Proc. XII Int. Congr. Genetics, Vol. 3: 349–370.

Allard, R. W., and Julian Adams. 1969b. Amer. Nat. 103: 621–645.

Allard, R. W., and A. D. Bradshaw. 1964. Crop Sci. 4: 503–508.

Allard, R. W., and S. K. Jain. 1962. Evolution 16: 90–101.

Allard, R. W., James Harding, and Conrad Wehrhahn. 1966. Heredity 21: 547–564.

Allard, R. W., S. K. Jain, and P. L. Workman. 1968. Advances in Genetics 14: 55–131.

Allard, R. W., Qifa Zhang, M. A. Saghai Maroof, and O. M. Muona. 1992. Genetics 131: 957–969.

Allard, R. W., P. Garcia, L. E. Sáenz de Miera, and M. Perez de la Vega. 1993. Genetics 135: 1125–1139.

Basnizki, J., and D. Zohary. 1987. Hort. Sci. 22: 678–679.

Basnizki, J., and D. Zohary. 1994. pp. 253–269. In: J. Janick (ed.), Plant breeding reviews, Vol. 12. John Wiley & Sons, New York.

Bateson, W. 1909. Mendel's principles of heredity. Cambridge.

Beal, W. J. 1876–1882. Report Michigan State Board of Agriculture.

Borlaug, N. E. 1954. Report First Int. Wheat Conference.

Borlaug, N. E. 1957. Report Third Int. Wheat Conference.

Borlaug, N. E. 1995. Wheat Breeding at CIMMYT. Wheat Special Report. No. 29.

Brackman, A. C. 1980. A delicate arrangement. Times Books, New York.

Braun, H. J., W. H. Pfeiffer, and W. G. Pollmer. 1992. Crop Sci. 32: 1420–1427.

Briggs, F. N. 1938. Amer. Nat. 72: 285–292.

Brown, A. H. D. 1975. Biometrics 31: 145–160.

Brown, A. H. D., and R. W. Allard. 1970. Genetics 66: 135–145.

Brown, A. H. D., and J. Daniels. 1973. Expl. Agric. 9: 321–328.

Brown, A. H. D., and J. Munday. 1982. Genetica 58: 85–96.

Browne, Janet. 1995. Charles Darwin: Voyaging, Vol. 1. Alfred A. Knopf, New York.

Bruce, A. B. 1910. Science 32: 627–628.

Castle, W. E. 1903. Proc. Amer. Acad. Arts and Sci. 39: 223–242.

Ceccarelli, S. 1989. Euphytica 40: 197–205.

Ceccarelli, S. 1994. Euphytica 77: 205–219.

Ceccarelli, S. 1997. Adaptation to low/high input cultivation. pp. 225–236. In: P. M. A. Tigerstedt (ed.), Adaptation in plant breeding. Kluwer Academic Publishers, The Netherlands.

Ceccarelli, S., and S. Grando. 1987. Euphytica 57: 157–219.

Ceccarelli, S., and S. Grando. 1991. Euphytica 57: 207–219.

Ceccarelli, S., and S. Grando. 1993. pp. 533–537. In: D. R. Buxton et al. (eds.), Conventional plant breeding to molecular biology. International Crop Science I CSSA, Madison, WI.

Ceccarelli, S., S. Grando, and J. A. G. von Leur. 1987. Euphytica 36: 389–405.

Clegg, M. T., and R. W. Allard. 1972. Proc. Natl. Acad. Sci. U.S.A. 69: 1820–1824.

Collins, G. B., and R. J. Shepherd, eds. 1996. Engineering plants for commercial products and applications. Annals New York Acad. Sci., Vol. 729, pp. 176 ff.

Collins, G. N. 1921. Amer. Nat. 55: 116–133.

Correns, C. 1900. Mendel's Regel über das verhalten der Nachkommens draft der Rassenbastarde. Ber. Deutsch. Bot. Ges., 17: 158–168.

Cowan, C. W., and P. J. Watson, eds. 1992. In: The origins of agriculture. Smithsonian Institution Press, Washington and London. (Includes articles on the origins of plant cultivation in the Near East, Africa, Europe, eastern North America, desert borderlands of North America, Mesoamerica and Central America, South America and Asia).

Crow, J. F. 1952. pp. 282–287. In: Heterosis. Iowa State College Press.

Crow, J. F. 1984. Genetics 33: 477–487.

Darwin, Charles. 1859. The origin of species by means of natural selection or the preservation of favored races in the struggle for life. Amer. ed. New York, 1868.

Darwin, Charles. 1868. The variation of animals and plants under domestication. Amer. ed. New York, 1877.

Darwin, Charles. 1872. The descent of man and selection in relation to sex.

Darwin, Charles. 1876. The effects of cross and self fertilization in the vegetable kingdom. Amer. ed. New York, 1877.

Davenport, C. B. 1908. Science 28: 454–455.

de Candolle, Alphonse. 1866. Origin of cultivated plants, 2d ed. Hafner, New York. Translated from 1886 edition.

Dobzhansky, T., and B. Wallace. 1953. Proc. Natl. Acad. Sci. 39: 586–591.

Doebley, J. F., M. M. Goodman, and C. W. Stuber. 1984. Systematic Bot. 9: 205–218.

Doebley, J. F., M. M. Goodman, and C. W. Stuber. 1985. Amer. J. Bot. 72: 629–639.

Doebley, J. F., J. Wendel, J. S. C. Smith, C. W. Stuber, and M. M. Goodman. 1988. Econ. Bot. 42: 120–131.

East, E. M. 1907. Rept. Connecticut Agric. Exp. Stat. for 1907.

East, E. M. 1908. Rept. Connecticut Agric. Exp. Stat.: 491–428.

East, E. M. 1916. Genetics 1: 164–176.

East, E. M. 1936. Genetics 21: 375–397.

East, E. M., and D. F. Jones. 1920. Genetics 5: 543–610.

Epperson, B. K., and R. W. Allard. 1984. J. Hered. 75: 212–214.

Fasoulas, A. C., and R. W. Allard. 1962. Genetics 47: 899–907.

FAO, Rome. 1970. FAO, State of Food and Agriculture.

FAO, Rome. 1987. FAO Production Yearbook.

Fehr, W. R., and H. H. Hadley. 1980. Hybridization of crop plants (a compilation of methods of hybridizing crop plants). Amer. Soc. Agron. and Crop Sci. Soc. Amer., Madison, WI. 765 pp.

Fienberg, S. E. 1980. The analysis of cross-classified categorical data. M.I.T. Press, Cambridge, MA.

Fisher, R. A. 1918. Trans. Royal Soc. Edinburgh 52: 399–433.

Fisher, R. A. 1930. The genetic theory of natural selection. Clarendon Press, Oxford.

Francis, C. A. 1986. Multiple cropping systems. Macmillan, New York.

Fyfe, J. L., and N. T. J. Bailey. 1951. J. Agric. Science 41: 371–378.

Galton, F. 1889. Natural inheritance. London.

Garcia, P., F. J. Vences, M. Perez de la Vega, and R. W. Allard. 1989. Genetics 122: 687–694.

Gauch, H. G. 1988. Biometrics 44: 705–715.

Gauch, H. G., and R. W. Zobel. 1996. Crop Sci. 36: 838–843.

Goodman, M. M., and C. W. Stuber. 1980. Maydica 28: 169–187.

Green, M. M., and K. C. Green. 1949. Proc. Natl. Acad. Sci. U.S.A. 35: 586–591.

Guttierez, M. G., and G. F. Sprague. 1959. Genetics 44: 1075–1082.

Haldane, J. B. S. 1946. Ann. Eugen. 13: 197–205.

Hardy, G. H. 1908. Science 28: 49–50.

Harlan, H. V. 1957. One man's life with barley. Exposition Press, New York.

Harlan, H. V., and M. L. Martini. 1929. J. Amer. Soc. Agron. 21: 407–409.

Harlan, H. V., and M. L. Martini. 1938. J. Agric. Res. 57: 189–199.

Harlan, H. V., and M. N. Pope. 1922. J. Hered. 13: 319–322.

Harlan, J. R. 1992. Crops and man, 2d ed. American Society of Agronomy, Madison, WI.

Hayes, H. K., and R. J. Garber. 1919. J. Amer. Soc. Agron. 11: 309–318.

Hayman, B. I., and K. Mather. 1955. Biometrics 11: 69–82.

Hull, F. H. 1945. J. Amer. Soc. Agron. 37: 134–145.

Hull, F. H. 1952a. pp. 451–453. In: Heterosis. Iowa State College Press.

Hull, F. H. 1952b. J. Amer. Soc. Agron. 37: 134–135.

Hutchinson, E. S., A. Hakim-Elahi, R. D. Miller, and R. W. Allard. 1983a. J. Hered. 74: 325–330.

Hutchinson, E. S., S. C. Price, A. L. Kahler, M. I. Morris, and R. W. Allard. 1983b. J. Hered. 74: 381–383.

Huxley, J. 1942. Evolution: The modern synthesis. Allen and Unwin, London.

Immer, F. R. 1941. J. Amer. Soc. Agron. 33: 200–206.

Jain, S. K., and R. W. Allard. 1966. Genetics 53: 633–659.

Jenkins, M. T. 1935. Iowa State College J. Sci. 3: 429–450.

Jenkins, M. T. 1940. J. Amer. Soc. Agron. 32: 55–63.

Johannsen, W. L. 1903. Uber Erblichkeit in Populationen und in Reinen Leinen. Gustav Fisher, Jena.

Johannsen, W. L. 1913. Elemente der Exakten Erblichkeitslehre, 3d ed. Fischer, Jena.

Johanson, D. C., and B. Edgar. 1995. From lucy to language. Simon and Schuster, New York.

Jones, D. F. 1916. Science 43: 509–510.

Jones, D. F. 1917. Proc. Natl. Acad. Sci. U.S.A. 3: 310–312.

Jones, D. F. 1918. Connecticut Agric. Exp. Stat. Bul. 207.

Kahler, A. L., and R. W. Allard. 1981. Theor. and Appl. Genetics 59: 101–111.

Kahler, A. L., R. W. Allard, and R. D. Miller. 1984a. Genetics 106: 729–734.

Kahler, A. L., S. Heath-Pagliuso, and R. W. Allard. 1981. Crop Sci. 21: 536–540.

Kahler, A. L., C. O. Gardner, and R. W. Allard. 1984b. Crop Sci. 24: 350–354.

Kannenberg, L. W., and R. W. Allard. 1967. Evolution 21: 227–7240.

Kauffman, Stuart. 1995. At home in the universe. Oxford University Press, New York, Oxford.

Knoll, A. A. 1992. Science 256: 622–627.

Ladizinsky, G. 1973. Chromosoma 42: 105–110.

Lerner, I. M. 1950. Population genetics and animal breeding. Cambridge University Press.

Lerner, I. M. 1954a. Proc. 9th Int. Congr. Genetics, pp. 124–128.

Lerner, I. M. 1954b. Genetic homeostasis. Oliver and Boyd, Edinburgh.

Lewis, E. B. 1951. Cold Spring Harbor Symp. on Quant. Bio. 16: 159–174.

Lewis, E. B. 1957. Cold Spring Harbor Symp. in Quant. Biol. 16: 159–174.

Li, C. C. 1955. Population genetics. University of Chicago Press.

Lush, J. L. 1945. Animal breeding plans. Collegiate Press, Ames, IA.

Lyell, Charles. 1830–1833. Principles of geology (facsimile reprint). 3 vols. University of Chicago Press, 1991.

Malthus, T. R. 1798. An essay on the principle of population, as it affects the future improvement of society, 6th ed. Murray, London. 1826.

Mather, K. 1943. Biol. Rev. 18: 32–64.

Mather, K. 1949. Biometrical genetics. Dover, New York.

Maunder, A. B. 1994. pp. 147–169. In: H. T. Stalker and J. P. Murphy (eds.), *Plant breeding in the 1990s.* CAB International, Wallingford, Oxon, U.K.

Mayr, E. 1991. One long argument. Harvard University Press, Cambridge.

Mendel, G. J. 1865. Versuche uber Pflanzenhybriden. English translation by W. Bateson in Jour. Roy. Hort. Sci. 26: 1–32. 1901.

Mendel, G. J. 1869. Letter of April 15, 1869, to C. Correns. English translation in Genetics 35 (suppl.): 1–29.

Morell, V. 1995. Science 268: 1279.

Nilsson-Ehle, H. 1909. Lund Univ. Arsskrift. 5: 1–122.

Pécaut, P. 1993. pp. 737–746. In: G. Kallo and B. D. Bergh (eds.), Genetic improvements of vegetable crops. Pergamon, Oxford.

Pécaut, P. 1993. pp. 737–746. In: G. Kallo and B. D. Bergh (eds.), Genetic improvements of vegetable crops. Pergamon, Oxford.

Pécaut, P., and C. Foury. 1992. pp. 460–470. In: A. Gallais and H. Bannerot (eds.), Amélioration des especés végétables cultivées. INRA, Paris.

Perez de la Vega, M. 1996. Plant genetic adaptedness to climatic and edaphic conditions. pp. 27–38. In: P. M. A. Tigerstedt (ed.), Adaptation in plant breeding. Kluwer Academic Publishers, The Netherlands.

Perez de la Vega, M., L. E. Sáenz de Miera, and R. W. Allard. 1991. Theor. Appl. Genetics 88: 56–64.

Perez de la Vega, M., P. Garcia, and R. W. Allard. 1991 Proc. Natl. Acad. Sci. U.S.A. 88: 1202–1206.

Pimbert, M. P. 1994. Seedling (July 1944): 20–25.

Principe, J. A. 1984. Hort. Sci. 19: 864–865.

Roberts, H. F. 1929. Plant hybridization before Mendel. Princeton University Press, Princeton, NJ.

Saghai Maroof, M. A., R. W. Allard, and Qifa Zhang. 1990. Proc. Natl. Acad. Sci. U.S.A. 87: 8496 ff.

Saghai Maroof, M. A., Qifa Zhang, D. B. Neale, and R. W. Allard. 1992. Genetics 131: 225–231.

Saghai Maroof, M. A., K. M. Soliman, R. A. Jorgensen, and R. W. Allard. 1984. Proc. Natl. Acad. Sci. U.S.A. 81: 8014–8018.

Sandison, A. 1959. Nature 184: 834.

Schwartz, D. 1960. Proc. Natl. Acad. Sci. U.S.A. 57: 1202–1206.

Shaw, D. V., A. L. Kahler, and R. W. Allard. 1981. Proc. Natl. Acad. Sci. U.S.A. 78: 1298–1302.

Shull, G. H. 1908. Rept. Amer. Breeders Assn. 4: 51–59.

Shull, G. H. 1909. Rept. Amer. Breeders Assn. 5: 296–301.

Shull, G. H. 1952. pp. 14–48. In: Heterosis, Iowa State College Press.

Smith, B. D. 1989. Science 246: 1566–1571.

Sprague, G. F., and J. F. Schuler. 1942. J. Amer. Soc. Agron. 24: 923–932.

Sprague, G. F., and L. S. Tatum. 1942. J. Amer. Soc. Agron. 34: 923–932.

Sprague, G. F., and W. T. Federer. 1951. Agron. J. 43: 535–541.

Sprague, G. F., and J. F. Schuler. 1961. Genetics 46: 1713–1720.

Stadler, L. J. 1942. Spragg lectures in plant breeding. Michigan State University.

Stephens, J. C., F. R. Miller, and D. T. Rosenow. 1967. Crop Sci. 7: 396.

Suneson, C. A. 1949a. Agron. J. 41: 459–461.

Suneson, C. A. 1949b. Agron. J. 48: 188–191.

Troyer, A. F. 1990. Personal communication.

Troyer, A. F. 1997. Breeding widely adapted, popular maize hybrids. pp. 185–196. In: P. M. A. Tigerstedt (ed.), Adaption in plant breeding. Kluwer Academic Publishers, The Netherlands.

Troyer, A. F., and W. L. Brown. 1976. Crop Sci. 16: 767–772.

Tschermak, E. 1900. Zeit. Landivirtsch. Nersuchwesen in Oestrruch.

Turesson, G. 1922. Hereditas 3: 211–221, 341–348.

Turner, J. R. G. 1967. Amer. Nat. 107: 195–221.

Vavilov, N. I. 1926. Studies on the origin of cultivated plants. Inst. Appl. Bot. Plant Breeding. Leningrad.

Vilmorin, L. de. 1856. Vilmoran-Andrieux, Paris.

Vries, H. de 1900. Ber. Deutsch. Bot. Ges. 18: 83–90.

Webster, O. I. 1976. Crop Sci. 16: 553–556.

Weisman, A. 1882. Studies in the theory of descent. English trans. by R. Mendola. 1875, 1876. Sampson Low et al., London.

Weismann, A. 1904. Die Selektimstheorie: Eine Untersuchung Jena: G. Fischer. (The Selection Theory). In: A. C. Seward (ed.), Darwin and Modern Science. Cambridge University Press.

Weismann, W. 1908. Jahreshafte Verein 64: 368–382.

Weir, B. S., and C. C. Cockerham. 1973. Genet. Res. 21: 247–262.

Westergaard, M. 1958. Advances in Genetics 9: 217–281.

Wentz, J. B., and S. F. Goodsell. 1929. J. Agric. Research 38: 505–510.

Wiltzen, E., and G. Fischbeck. 1990. Plant Breeding 104: 58–67.

Woodward, C. M. 1931. Purnell Corn Imp. Rept., pp. 48–49.

Wright, S. 1921. Systems of mating. Genetics 6: 111–178.

Wright, S. 1937. Proc. Natl. Acad. Sci. U.S.A. 23: 307–320.

Wright, S. 1949. Proc. Amer. Phil. Soc. 93: 4471–4478.

Wright, S. 1968. Evolution and Genetics of Populations. Univ. of Chicago Press. Vol. 1.

Yearbooks of Agriculture. 1936, 1937. Give detailed descriptions of floral structure and mode of reproduction of many cultivated plants.

Yu, S. B., J. L. Li, C. G. Xu, Y. F. Tan, Y. J. Gap, X. H. Li, Qifa Zhang, and M. A. Saghai Maroof. 1997. Proc. Natl. Acad. Sci. U.S.A. 94: 9226–9231.

Yule, G. V. 1906. Rept. Third Int. Cong. Gen. pp. 1409–1422.

Zhang, Qifa, M. A. Saghai Maroof, and R. W. Allard. 1990. Theor. Appl. Genetics 80: 121–128.

Zhang, Qifa, X. K. Shen, M. Dai, M. A. Saghai Maroof, and Z. B. Li. 1994. Proc. Natl. Acad. Sci. U.S.A. 9: 8675–8679.

Zohary, D., and M. Hopf. 1988. Domestication of plants in the old world. Clarendon Press, Oxford.

# Index

Lightning Source UK Ltd.
Milton Keynes UK
UKOW06n1656080515

251172UK00001B/41/P